T0073969

Palgrave Advances in the Economics of Innovation and Technology

Series Editor
Albert N. Link
University of North Carolina at Greensboro
USA

The focus of this series is on scholarly inquiry into the economic foundations of technologies and the market and social consequences of subsequent innovations. While much has been written about technology and innovation policy, as well as about the macroeconomic impacts of technology on economic growth and development, there remains a gap in our understanding of the processes through which R&D funding leads to successful (and unsuccessful) technologies, how technologies enter the market place, and factors associated with the market success (or lack of success) of new technologies.

This series considers original research into these issues. The scope of such research includes in-depth case studies; cross-sectional and longitudinal empirical investigations using project, firm, industry, public agency, and national data; comparative studies across related technologies; diffusion studies of successful and unsuccessful innovations; and evaluation studies of the economic returns associated with public investments in the development of new technologies.

More information about this series at
http://www.springer.com/series/14716

Swati Bhatt

How Digital Communication Technology Shapes Markets

Redefining Competition, Building Cooperation

palgrave
macmillan

Swati Bhatt
Department of Economics
Princeton University
Princeton, New Jersey, USA

Palgrave Advances in the Economics of Innovation and Technology
ISBN 978-3-319-47249-2   ISBN 978-3-319-47250-8 (eBook)
DOI 10.1007/978-3-319-47250-8

Library of Congress Control Number: 2016954957

Cover illustration: Détail de la Tour Eiffel © nemesis2207/Fotolia.co.uk

Printed on acid-free paper

This Palgrave Macmillan imprint is published by Springer Nature
The registered company is Springer International Publishing AG
The registered company address is: Gewerbestrasse 11, 6330 Cham, Switzerland

To Ishaan
For whom connectivity means so much more than technology

PREFACE

This book is written for those who are looking for a way of conceptualizing the economic impact of the Internet. It does not focus on a particular problem nor is it aimed at my peers at academic institutions. If you have been wondering about how the various apps you use, the websites you browse, the text messages you use to communicate fit into some larger worldview then you may find some answers in this book. I aim to provide a framework for thinking about the Internet so that new developments or variations on existing products and services can be contextualized within this construction.

The process of writing this book began in the spring of 2011 when Dan Mavraides – one of my senior thesis advisees at Princeton, and captain of the very successful Princeton University men's basketball team – walked into my office, three weeks before the thesis deadline. After having led the team in March Madness or the NCAA Division I Men's Basketball Tournament, he was diligently polishing his thesis. In the course of our discussion, he asked a great question. In the age of smartphones, will more fans attend basketball games in person or will they simply view them on their devices? Questions such as these fueled my research and prompted me to design and teach a course on the "Economics of the Internet" in the fall of 2013, and then every subsequent fall. The interactive nature of this course triggered fascinating discussions with my students, many of whom subsequently wrote their senior thesis or junior research paper on the subject. I have had the privilege of supervising over 150 senior theses during the past quarter century at Princeton University.

It has been a joy to interact with many bright and enthusiastic students over the years. This new digital world is their world and there is much I have learnt from them; the names I thank below constitute just the tip of the iceberg.

Samvitha Ram and Pia Sur were superlative critics of my early drafts. They took my class and started asking profound questions before the end of the first week. Darwin Li, my advisee for the past four years (including my class, junior and senior thesis), provided excellent editorial assistance and diligently conducted the empirical research for the diagrams in Chap. 3.

Jason Yu (also in my class, and advisee for four years) did his senior thesis on Uber, challenging my thinking on labor markets. Chuck Dibilio and Florin Radu, sitting in the back of the class, always extended the discussion in novel ways. Hillary Bernstein, ever engaged and sitting in the front of the class, kept me alert and on target. Dalia Katan reminded me that advertising, correctly, is at the heart of digital business models. Jose De Alba and Tom Pham taught me how to "Venmo" as college students do it, and Hannelora Everett made me understand, at a personal level, the security added by the Red Alert app in Israel. Linda Zhong made me reevaluate my thinking about online courses, arguing that they can never capture the professor–student dynamics of a classroom; Lillian Cartwright gave a face to technology in talking about her work with GiveDirectly, a non-profit in Kenya, that targets thatched roofs using Google Earth technology to hand cash directly to the impoverished; and Garrett Frey taught me about the potency of Instagram's multi-interface sharing capabilities.

Conversations in my office with students were another source of stimulation and delight, and the names that follow are only a partial list: Anna Matlin, Haebin Kim, Karina Marvin, Saahil Madge, Kenny Anhalt, Yasin Hegazy, Eric Rehe, Ryan Albert and Everett Price. Spontaneous in-class debates between Judy Hou, Kevin Tang, Ambika Vora and Charles Zhou were an affirmation of the excitement of the general topic.

I am deeply appreciative of the support and stimulation offered by my friends in the Economics Department at Princeton, many of whom I have known since I was a graduate student: Avinash Dixit, Gene Grossman, Alan Blinder, Harvey Rosen, Alan Krueger, Dilip Abreu, Anne Case and Janet Currie. I have been inspired by the seminars and conversations with faculty at Center for Information Technology Policy at Princeton: Brian Kernighan, Ed Felten, Nick Feamster, Paul DiMaggio and Zeynep Tufekci.

My editors at Palgrave Macmillan, Allison Neuburger and Sarah Lawrence, were extremely patient, generous with their advice and provided the camaraderie that makes endeavors such as this rewarding.

My own family network is a treasure. My daughter Andie, with whom three-hour brainstorming sessions across the continent are a unique source of joy and inspiration, has been my closest advisor and best friend. My dear husband, Ravin, who has been my intellectual Google ever since he helped me with Brownian motion concepts when I was struggling with options pricing formulae in graduate school, has been my deepest connection. My son Ishaan, the wisest treasure, whose autism and struggle with connections has made me truly understand what connectivity is all about.

CONTENTS

1 The Technology: Has the Digital Communication
Technology Changed the Way Markets Function?
Cooperation or Competition?   1

2 The Three Drivers: Connectivity, Data and Attention   17

3 The Three Trends: Granularity, Behemoths and
Cooperation   29

4 The Independent Contractor and Entrepreneurship
in Labor Markets   57

5 The On-Demand Economy and How We Live:
Communication, Information, Media and Entertainment   71

6 The Sharing Economy: Information Cascades, Network
Effects and Power Laws   105

7 The Private World of Sharing and Cooperation   119

8 The Internet and Regulation   133

9 The Conclusion 143

Index 149

LIST OF ABBREVIATIONS AND ACRONYMS

Connections (n.) –	Both the individuals (nodes) who are connected and the means (links) by which they are connected
Connectivity (n.) –	The state of being connected
Digital communication technology (DCT) –	Technology that transmits information in digital form
Digital economy –	An economy empowered by DCT
FOMO –	Fear of missing out, leading to the phenomenon of copying
Firm, age –	Measured in number of years in existence
Firm, size –	Measured in number of employees
Intermediary, passive –	Requires no deep knowledge of the characteristics or motives of the various economic agents who participate in a given market
Matching market –	Markets that require search for the perfect counterparty for a given transaction
Network economy –	Connections between economic agents
Network effects –	An increase in the value of a product when the size of the network of users expands

Platform –	Passive, online intermediary, a bazaar for buyers and sellers, who originate and consummate the transaction themselves
Private equity –	An earlier stage of venture funding, associated with active engagement in the businesses
Social media –	Platforms that manage the network of individuals, and on which communication takes place
Social network –	Network of individuals who share and communicate on the social media platform and produce content
TRR&R –	Trust, reputation, responsibility and rights forming the basis of social capital
Venture apital –	Funds provided by a group of private investors who invest their own money into young, entrepreneurial ventures

LIST OF FIGURES

Fig. 1.1   (a) Path; graph not connected. (b) Cycle; directed graph;
           connected graph. (c) Two distinct components; each is k
           connected cluster; structural hole between them. (d) $\underline{G}$ is
           gatekeeper; $\underline{G}$ is pivotal from path $\underline{J}$ to B; GA is a bridge;
           $NO_{AG} = 0$ (neighborhood overlap = 0). (e) $CC_A = 0$ $CC_A = 1$;
           $NO_{AG} = 0$; $NO_{AG} = 1$. (f) $CC_A > 0$; $NO_{AG} = 0$. (g) Buyers;
           traders; sellers; [B, C]; [A, G]; [J, K]   13
Fig. 3.1   Small firms as a percent of total firms   38
Fig. 3.2   Number of employees   39
Fig. 3.3   Number of seed investment deals   41
Fig. 3.4   Growth of VC funding   42
Fig. 3.5   Number of US tech IPOs   43
Fig. 3.6   Number of US tech private deals   45
Fig. 3.7   Amount (USD) of US tech private funding   46
Fig. 5.1   The digital entertainment value chain   77
Fig. 5.2   Single and multi-homing for users   83

CHAPTER 1

The Technology: Has the Digital Communication Technology Changed the Way Markets Function? Cooperation or Competition?

Abstract As social beings, humans have traded goods and services for generations. The mechanics of exchange have adapted over time to the changing environment but the underlying incentives remain the same. Differences in preferences and resources are bridged by connecting with individuals in other parts of the economic network using digital communication technology (DCT). DCT has generated new scarcities and new mechanisms for resource allocation, enlarging the scope of exchange along three dimensions. One is the trend toward organizational restructuring (OR) precipitated by granularity and disintermediation. The second trend is the emergence of organizational behemoths (OB), fostered by network effects. And third, competition for scarce resources is channeled into cooperation as individuals adapt to a rapidly changing context for exchange, and this adaptation itself changes the environment.

Keywords Granularity · Organizational restructuring (OR) · Disintermediation · Organizational behemoths (OB) · Competition as information

Economic transactions between individuals have existed since ancient times. They have evolved and adapted to changing circumstances, such as climate change, epidemics, famine, and human nature, periodically creating imbalances in the economic system. Markets, as a mechanism for scarce

S. Bhatt, *How Digital Communication Technology Shapes Markets*,
Palgrave Advances in the Economics of Innovation and Technology,
DOI 10.1007/978-3-319-47250-8_1

resource allocation under competing demands, have adapted to these imbalances. Currently, we are in the throes of another shock to the system, a technological shock in the form of digital connectivity, creating a network of economic agents, a network economy. What does this perturbation mean for resources and trade? Are there new scarcities? Is the exchange of goods and services governed by new rules? How can we characterize the network economy?[1] If the rules of the trading game have changed, is there more cooperation or competition?

My answer, in brief, is that we are witnessing new scarcities and new mechanisms for resource allocation, as digital communication technology (DCT) shifts boundaries between economic agents. The scope of trade has been redefined along three dimensions. One is the trend toward organizational restructuring (OR) leading to granularity and disintermediation on an economy-wide scale. The second trend is the development of organizational behemoths (OB). And the third trend is a recalibration of competition.

The multiple demands of information and communication have shifted the boundaries between public and private goods and struck against the hard resource constraint of attention. Consequently, a new scarcity has surfaced in the form of cognitive bandwidth, or more simply attention. Attention is a scarce resource and, therefore, a tradable commodity in the world of marketing. Individuals provide personal information, at no cost, in exchange for freebies. This data is sold to marketing firms who create precisely tuned messages and purchase advertising space to acquire consumer attention or eyeballs. Individuals' attention is effectively traded when this personal data is sold and then sliced and diced so as to best capture "brain space."

In the restructuring trend (OR), products are becoming more granular and the elimination of intermediaries is making large organizations less dominant in the economic landscape.[2] Granularity of products arises from an unbundling of lumpy purchases, such as ridesharing replacing vehicle ownership. Organizational granularity empowers smaller trading units so that control is retained with a concomitant reduction in the size of these trading units, where size is measured by number of employees. Connectivity allows information transfer, which enables restructuring the organization of trading units into smaller entities. These granular person-to-person (P2P) trading units retain authority and autonomy. Consumer demand for functionality has replaced purchases based on brand or product name as buyers use iPads, for example, for purposes beyond their original intention. This fosters innovative product differentiation by projecting user behavior over time such that newer products and services can cross traditional boundaries. More fundamentally, connectivity generates heterogeneity of ideas and cultural diversity, providing fertile ground for startups and entrepreneurial activity.

The number of small firms in the US has been increasing year over year since 2005, where small is defined as firms that have between 3 and 250 employees. But while the number of small firms is increasing, the number of employees at these firms is decreasing. Entities such as Elance ODesk (founded in 2003, but rebranded as Upwork in 2015), Grubhub (2004), Airbnb (2007), TaskRabbit (2008), Uber (2009), Blue Apron (2012), Instacart (2012), and Shyp (2014), were all founded as small firms. While firm age and size are likely to be correlated at inception, the notion of startups is best captured by firm age and not firm size. The data on entrepreneurial activity suggest a resurgence of entrepreneurship: The Kauffman Startup Activity Index, a composite of entrepreneurial activity in the US, increased in 2015, reversing a five-year downward trend that began in 2010. Particularly noteworthy is that this index defines startups as firms younger than one year, with at least one employee other than the owner.[3]

Most measures of digitization of the economy focus on the *implementation* of digital technology: investment in digital assets, access to broadband and mobile devices, and the incorporation of this technology into the production process. More difficult to measure is the set of economic possibilities – in terms of new products, new markets, and new ways of doing business and consuming – arising from this technology. Disintermediation and granularity in firm size has fostered a form of market design based on cooperation among trading units. Cooperation and coordination are inherent to market economies where all entities play by the same rules and where prices serve as the hand of coordination, matching demand with supply. But cooperation in the context discussed here is beyond observing the rules of the game – it rests on the sharing of information. This feature is not due to altruism but rather the best response to the unfathomable forces of connectivity.

The second trend, development of OB, arises as connections explode across the global network of individuals and information advantages are overwhelmed by network size effects. OR in some industries, that would favor granularity, is upended in favor of "too big to fail" trading units. The digital economy is a network economy much like the aspen root system. For every aspen grove you see above ground, there is a vast system of underground roots connecting multiple groves. The roots can get entangled and the aspen trees can get denser within a single grove. Similarly, as connections multiply, powerful network effects, information cascades and data-driven algorithms lead to concentration of economic power in OB. These hubs foster homogeneity of ideas, further enhancing benefits of like-minded connections in a reinforcing feedback loop.

We are familiar with Amazon dominating in the retail arena, Facebook in the social dimension, Google in the search area and Netflix in entertainment. If we add Microsoft, Ebay, Priceline, Salesforce, and Starbucks, we have a network economy that is culturally homogenous and short of business dynamism.[4] The risk of starting new ventures increases in this hyper-connected network. Startups risk obliteration under the glare of "either-you-are-with-us-or-against-us" economic thinking or are acquired by the behemoths, and then torn apart, absorbing only those pieces that add synergy to the acquiring firm. Innovative thinking requires solitude, so hyper-connectivity may play a role in the apparent shortage of business churning or dynamism. Steven Spielberg said in his Harvard commencement speech, "Social media that we're inundated and swarmed with is about the here and how... this is why it's so important to listen to your internal whisper" [2]. With cell phone chimes smothering the tender shoots of breakthrough ideas, we might remain in the "too-big-to-fail" grove.

I will be discussing these ideas in the chapters that follow, but for now, let me return to the traditional models. In order to build the connection between technology and disintermediation, we have to start with the most elementary feature of economies – markets.

Trading in markets dates back to the days when farming replaced foraging around 7000 BCE in China and Mesopotamia, what is today Iraq, Jordan, and Syria. Later, around 2200 BCE, we have records suggesting that Egypt's pharaohs, rich in gold and grain, gave these goods to minor rulers of Lebanese cities, who reciprocated with fragrant cedar. According to Ian Morris [3], "gift exchange was as much rooted in psychology and status anxiety as in economics, but it moved goods, people and ideas around quite effectively." Whether the bilateral exchange of goods and services was motivated by considerations of common economic advantage or psychological well-being, barter-trade was recorded somewhere between the seventh- and fifth-century B.C.E when Joseph (of Biblical fame) was sent by his family to Egypt to exchange his coat for food. Prior to that time, the Xia dynasty ruled in Anyang in the Yellow River Valley in China somewhere between 2000 and 1600 BCE, where evidence of trade is depicted in the third-century CE novel of Luo Guanzhong, "The Romance of the Three Kingdoms" [4]. "And the three brothers went forth to welcome the merchants. They went northwards every year to buy horses . . . and gave the brothers fifty good steeds, and besides, five hundred ounces of gold and silver." Importantly, according to Joseph Schumpeter's analysis, Aristotle was already writing in 300 BCE about the requirement of "money" as a medium of exchange, unit of account and store of value, and that money must itself be one of the commodities exchanged. So the world had moved straight from barter to the use of a medium of exchange in markets by 300 BCE [5].

The design of markets has been stable at least since the time of Aristotle. In ancient Greece, these bazaars were physical congregations of buyers and sellers. But it was not as simple as that. The product quality was uncertain and the exchange rate, or "equivalence" in Aristotle's terminology, was ambiguous.[5] What was the precise composition of tin and copper in the bronze weapon that was for sale? How were you to believe the seller? Were there large inventories of this material so that sellers were eager to dispose of their wares? Did this depress the exchange rate between weapons and horses, for example? Did different sellers have different-sized inventories and if so were buyers aware that some sellers might be willing to bargain for lower values? Did buyers have to search the entire bazaar for these "lower-value" sellers and if so were these search costs manageable? As John McMillan [6] writes, "Two kinds of market frictions arise from the uneven supply of information. There are search costs: the time, effort and money spent learning what is available, where and for how much. And there are evaluation costs, arising from the difficulties buyers have in assessing quality. A successful market has mechanisms that hold down the costs of transacting that come from the dispersion of information."

In modern times, even with clearly stated prices and observable quality, we have versions of barter where the seller might deliberately obfuscate the exchange rate. For example, many years ago as a teenager visiting my aunt in Baroda, India, I observed her buy vegetables from the local vegetable cart that made morning rounds in the neighborhood. While the central market price for eggplants was around INR 3 per kilo, the vendor, more like a family member who received repeat business and unlikely to deceive, charged slightly more. Instead, he would throw a couple of green chilies and a few pieces of ginger into the bag of eggplants as a "bonus." The vendor obtained his desired price and my aunt gleefully thought she got a bargain. The exact exchange rate or price remained ambiguous despite numerous neighboring vendors and low search costs.

While the story above may sound apocryphal, repeated interaction between the same two parties plays a powerful role in some markets. The situation, however, is not universal, since in general, price transparency is needed for markets to function efficiently. The actual mechanics of price transparency is via information, not some complicated system. Information is revealed about (i) what is available, (ii) where it is available, and (iii) at what exchange rate it is available, thus fulfilling the matching and price discovery functions of markets. But what are the pre-conditions for prices to provide accurate information?

Competition in the traditional neoclassical economic model is defined as an environment with multiple traders, whose interaction agglomerates disparate bits of relevant information, making prices accurate representations of the underlying technology and tastes. When prices accurately convey information about the exchange value of the good, they provide stabilizing, self-correcting incentives to economic agents, who respond to these price signals by updating their purchase-and-sales decisions. In turn, prices themselves react to fundamental imbalances so the cycle is reinforcing. This dynamic represents competitive activity in markets – traders adapt and learn from interacting with each other and prices adjust along with quantities.

There are three perspectives of competition. First, an environment of multiple traders; second, an interactive process of adaptation and learning; and third, an outcome of informative prices.[6] What definition of competition is truly meaningful in a discussion of the network economy? The first aspect of the definition had more support in the pre-Internet days when the only way to have informative prices was the agglomeration of information from multiple traders, in a static model. Therefore, the second and third aspects – the process and outcome – are more relevant today.[7] To be clear, in the network economy, the primary mechanism for information sharing is connectivity itself, and not the resultant granularity. Connectivity *automatically* leads to informative prices. So, in a network economy, a meaningful definition of competition is informative prices, not necessarily multiple traders.

The question posed at the beginning of this chapter – does the Internet move markets toward more competition or more cooperation – is best answered by recognizing this definition of competition. In the network economy, information surges across the network and makes proprietary hoarding of information impossible. This means that the Internet is moving markets toward a new definition of competition since every transaction is based on knowledge sharing made cheap and fast by digital connectivity. Hence, granularity is the result of cooperative information gathering and sharing.

Two basic prerequisites of markets, writes Avinash Dixit, are "security of property and of contract" [7]. Meaningful economic exchange requires confidence in ownership rights over the property or good to be bought or sold and confidence in the successful implementation of the contract. The exchange must be carried out and reneging should be costly; hence legal contract enforcement is vital. In the network economy, where information is the key commodity, the notion of property rights over information becomes awkward to define. How can you restrict the flow of information when it is cheap or costs nothing to share this information between multiple economic units? Information as a product is non-excludable like oxygen in the atmosphere – once it is out there you cannot exclude people from consuming it. It is also non-rival, like viewing a sunset, since my consumption of information doesn't preclude anyone else from consuming this same information.

In the days of Luo Guanzhong, these informational requirements were addressed by the intervention of friendly and neutral traders, who became the intermediary between the seller and buyer of goods and also the monopoly holder of valuable information. These traders could be the horse dealers who traded horses for gold and steel weapons. Using reputation as collateral, these traders would assure both parties of the veracity of their proposed contract, thus ensuring the trade. Thus was born what became known as an intermediary-trader, who created and monitored the trading links in bazaars, managing trading orders by ensuring the successful matching of buyers and sellers. As John McMillan writes, "Market intermediaries like wholesalers and trading companies reduce search costs for firms" [6].

Moving from ancient bazaars to the Internet bazaar, the release of the iPhone in January 2007 marked the introduction of a technology that has connected or linked markets, so that information is instantly, continuously, and ubiquitously available to all participants. The intermediary-trader was eliminated. The phone was no longer merely a communication device, but a computer and a camera, linking the actual world to the virtual world. The smartphone changed the way people connect to each other and the world. Three key game changers or major drivers of the network economy had been unleashed: mobile connectivity, a vast collection of facts or big data, and a new concept of time.

Connectivity affects the economics of networked markets by eliminating intermediaries, so might the Internet possibly be driving the fourth industrial revolution (following the steam engine; the systematic application of science to technology as in the internal combustion engine, electricity, telephone, telegraph; and the retail productivity revolution as manifested in the assembly line)? Disintermediation in financial markets is evidenced by innovations such as the digital wallet (Apple Pay, Google Pay, Venmo, and Square), shortening the financial supply chain by unbundling the intermediation function of bank deposits and providing a payment mechanism, but not a store of value. Boundaries between products and between industries have become blurred giving rise to a fluid economy where definitions of markets and transactions are ambiguous. For example, the distinction between employee and contract labor is becoming blurry. Facebook is now giving its contract workers more benefits such as paid leave, which makes them more like salaried employees [8]. Connectivity also encompasses organizational change. Industries such as education, healthcare, and the news media were already highly connected, but this new technology has reconfigured the map of these industries. Education must now be redefined to include the virtual classroom or massive, open, online courses (MOOCs); healthcare must include customer review-driven online health newsletters and news media must include social networks such as Twitter, which deliver news in real time. Not only are there new industry definitions, but new uses for old products. The iPhone is a voice communication device and a camera, both of which existed a decade ago, but blending these two capabilities not only defines a new use for old products but redefines the product itself.

Connectivity becomes more potent in its impact because of the mobile dimension – the anytime and anywhere idea. To put mobile connectivity into perspective, consider some numbers. According to a recent Pew Survey, over two-thirds (64 %) of American households own a smartphone and 7 % of American households do not have any Internet connectivity, at home or elsewhere, other than via their smartphones. This smartphone dependency is income related since only 1 % of households earning more than $75,000 are so deeply reliant on the device for Internet connectivity. Of those that own smartphones, 68 % used their phone for breaking news, 62 % used the device for health information, and 57 % for mobile banking. The smartphone has become a necessity for the 46 % of smartphone owners who say they cannot live without it [9].

Painting an extreme scenario, Jeremy Rifkin makes the case that connectivity will undermine the capitalist economy by driving production costs to zero and will be replaced by the Collaborative Commons. The economy he envisions is one where ownership and exchange is replaced by communal sharing. Social capital becomes the bedrock of this economy, where connectivity "brings the human race out of the age of privacy, a defining characteristic of modernity, and into the era of transparency." Furthermore, he adds, "While privacy has long been considered a fundamental right, it has never been an inherent right." In fact, the worst punishment meted out to members of society is ostracism. [10]

Currently, we are only witnessing the game-changing role played by mobile connectivity. We have yet to feel the impact of big data. Big data, writes Steve Lohr, is a "catchphrase" which

> stands for the modern abundance of digital data from many sources – the web, sensors, smartphones and corporate databases – that can be mined with clever software for discoveries and insights. Its promise is smarter, data-driven decision making in every field. [11]

For example, the growing field of passive telemetrics enables the gathering of unprecedented amounts of longitudinal data that assists in behavioral assessment for monitoring and improved understanding of individuals with autism spectrum disorders (ASD). Matthew Goodwin [12] writes, "[W]e are developing passive telemetric audio and video technologies for densely sampling behavioral manifestations of ASD in the first two years of life," which is when most behavioral abnormalities are evident but not diagnosed. Early, prescient diagnosis enables effective, long-term treatment.

Human interaction and engagement generate an intricate web of connections. These connections form coherent patterns that determine how information is transmitted through the network of relationships. Simultaneously, the very *exchange* of ideas across the network – the transmission of information – determines human behavior and network formation. For example, strong social ties mobilize individuals to act in a social network, whereas economic incentives, which focus on the rational individual, may fail to generate action. Social connections provide value to individuals and these connections can be used to apply pressure for change. In a seminal article, Matthew Jackson writes, "Networks are not only conduits for information or influence, but also adjust in reaction to behaviors" [13].

These ideas are elaborated upon according to the following map. After a brief explanation of the language of networks in the appendix to this chapter, Chap. 2 explains *why* connectivity, data and a new resource scarcity, attention, are the most significant forces of DCT. In Chap. 3, I explain *how* these drivers lead to disintermediation, granularity, and cooperation by applying network theory to economic relationships in order to characterize the network economy. The position of an economic unit or node within the network impacts the nature of links between nodes, so the structure of the network has powerful effects on the outcome in terms of prices and welfare.

It has been argued that four industries have powerful effects across the economy. In the language of networks, these industries have a centrality that is not shared by other industries. In this book, I will discuss the education and entertainment industries, leaving the other two industries, energy and environment, for future exploration. Accordingly, I will examine the impact of network structure on education and labor markets in Chap. 4, and the world of entertainment in Chap. 5, captured more broadly in the CIME industries – Communication, Information services, Media (publishing including software, books, newspapers), and Entertainment (motion pictures, sound recording, broadcasting). I broaden the scope, in Chap. 6, to consider macro-level effects in financial markets – information cascades, power laws, and network effects. Chapter 7 considers the intersection of privacy (starting with the law encapsulated in the First and Fourth Amendments to the US Constitution) and technology. The intersection of commerce and regulation that is influenced by, and in turn impacts, the architecture of the Internet is discussed in Chap. 8. Cooperation versus conflict in the network economy is best understood in this framework. Self-determination, liberty, and the freedom of ideas are caught up in the struggle for control over this Internet. Finally, I conclude with some thoughts about why cooperation is driven, not simply by altruism, but by rational calculation.

Let me clarify two important points of nomenclature. First, I use the acronym *Internet and DCT* to refer to (i) the *architecture* of mobile, DCT, that is, a network of distributed computing systems and (ii) the entire *cyberspace* of connectivity that transmits information in digital form. DCT includes artificial intelligence and the Internet of Things, which is connectivity across inanimate objects. Second, I use the term network economy to capture connections between economic agents and the term digital economy to signify that these connections are digitally powered. The two are inextricably related and I use them interchangeably, depending upon context.

My Take[8]

DCT has shifted the boundaries between economic agents and between public and private goods. Boundaries between firms are reorganized, as firms are restructured into granular entities as well as behemoths. Information transparency has moved the needle on privacy and encroached upon individual autonomy, so we fight for control over personal data – our attention, our work, and our access to information. But in order to adapt to the changing and unfathomable environment, our best response is to cooperate when we cannot control.

Appendix – Graph Theory

The Internet means connectivity, so we need a clear definition of connectedness. Graph theory is a branch of mathematics which provides us with a framework for understanding connectedness [14].[9]

A graph is a representation of a network. It consists of nodes which are connected by links or edges. Nodes represent market participants or economic agents and links represent interactions between them. These interactions could be actual transactions or simply trading arrangements, where some form of economic interaction is, or will be, consummated.

These links are of two types. The first is a link characterized by the logical structure of the network, the basic pattern. This pattern is governed by the logic of belonging to a particular group, a market where nodes trade. The second is a link created by the fact that individual outcomes are impacted by the behavior of all other individuals in the network. In this way, links are governed by behavior.

Links can be formed between nodes because they have a mutual friend (Triadic Closure), they share a common interest (Focal Closure), or they influence friends to join an interest (Membership Closure).

There are many patterns of connections between nodes. A *path*, Fig. 1.1(a), is a sequence of linked nodes. A *cycle*, Fig. 1.1(b), is a path where the initial and terminal nodes are identical. Figure 1.1(b) is also a *directed graph*, where every link has directionality and points from some node A to some other node B, and is also a *connected graph*, since there is a path between every pair of nodes. A graph is strongly connected if there is an unbroken path between every pair of nodes.

The *distance* between nodes is the shortest number of hops or links between these nodes.

Components are distinct groups of linked nodes and a *cluster* is a component with densely connected nodes, as in Fig. 1.1(c). A *structural hole* is the empty space between two components that may be linked via a bridge.

A *gatekeeper node*, G, in Fig. 1.1(d) has the property that all nodes in one component need to go through this gatekeeper to get to other components. Consequently, gatekeepers have enormous power over components. Copyrighted technology, such as computer software, can have greater economic significance than other copyrighted material such as books and movies. Computer software often controls the gateway or interface between other software and the copyrighted software in question. Microsoft has copyright protection over the interface between the Windows operating system for desktops and operating systems on servers. Some would argue that search engines, such as Google, have similar gatekeeper status granted by consensus across users.

Node G is *pivotal* since the shortest path between two components goes through this node.

G-A is a *bridge* since it connects distinct components and is the only path between these two components; a local bridge is the shortest path between two nodes but not the only path. Since the end nodes of a bridge have gatekeeper status, they have no common neighbors.

Some ratios reveal subtle underlying relationships such as the clustering coefficient of a node A (CC_A) and neighborhood overlap of a link AC (NO_{AG}). Both CC_A and NO_{AG} are between zero and one.

The *clustering coefficient* of node A, or CC_A, is the proportion of total neighbors that are linked. If many of your friends also know each other then you have a high clustering coefficient. For example, if you belong to a group, then many of your friends also belong to this group. Figure 1.1(e) gives examples of a zero clustering coefficient as well as a perfect clustering coefficient of one.

The *neighborhood overlap* of a link AG, or NO_{AG}, is the proportion of total neighbors of both A and G that are shared by both nodes. If there are many common friends between A and G, then there is high neighborhood overlap. The number of common friends is called the *embeddedness* of the link.

In Fig. 1.1(e), suppose A represents Apple on the left diagram and none of its developers know each other, then the clustering coefficient of Apple is zero. On the right diagram in Fig. 1.1(e), if the Apple Watch brings developers together, then Apple's clustering coefficient is one.

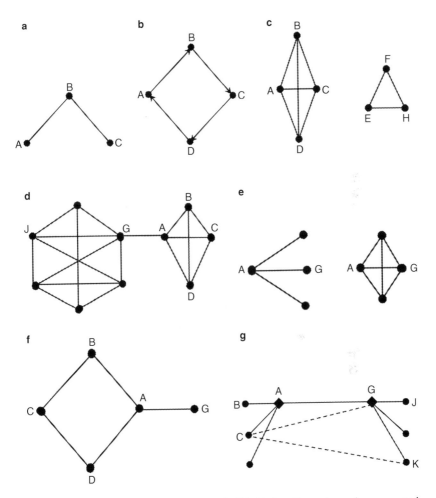

Fig. 1.1 (a) Path; graph not connected. (b) Cycle; directed graph; connected graph. (c) Two distinct components; each is k connected cluster; structural hole between them. (d) $\underline{G}$ is gatekeeper; $\underline{G}$ is pivotal from path $\underline{J}$ to $\underline{B}$; $\underline{GA}$ is a bridge; $NO_{AG} = 0$ (neighborhood overlap = $\overline{0}$). (e) $CC_A = 0$ $CC_A = 1$; $NO_{AG} = 0$; $NO_{AG} = 1$. (f) $CC_A > 0$; $NO_{AG} = 0$. (g) Buyers; traders; sellers; [B, C]; [A, G]; [J, K]

As another example, in Fig. 1.1(d), A and G are friends and A works at Amazon and G works at Google, but none of the other workers at either of these institutions know each other. Then the neighborhood overlap of the link AG is zero or $NO_{AG} = 0$. However, both A and G have bridging capital and have the opportunity to create potentially useful connections between these two distinct groups. Now, suppose there is a merger between Amazon and Google, and all colleagues of A and G get to know each other (not shown in diagram). When both A and G share all their neighbors, $NO_{AG} = 1$, thus generating bonding capital. We can also say that A and G are in a highly embedded network.

While the clustering coefficient is a powerful concept for nodes, neighborhood overlap addresses connectivity at the level of the network. The two ratios are independent of each other, except in one case. Consider a node A and some neighbor G in the left diagram in Fig. 1.1(e). If $CC_A = 0$ then $NO_{AG} = 0$ because if not then there exist shared neighbors, which implies that two of A's neighbors are linked which contradicts the assumption that $CC_A = 0$.

No such implications can be drawn when either ratio is 1 or when neighborhood overlap is zero. For example, in Fig. 1.1(d), $NO_{AG} = 0$ but $CC_A > 0$. This is because $CC_A > 0$ does not preclude $NO_{AG} = 0$. All of A's neighbors can be linked, except G, maintaining the idea that A and G share no neighbors.

Similarly, again in Fig. 1.1(d), $CC_B = 1$ does not imply that $NO_{AB} = 1$. While all of B's neighbors are linked, A's neighbor G is not linked to B. Hence, $NO_{AB} < 1$.

Finally, $NO_{AG} = 1$ does not imply that $CC_A = 1$, since shared neighbors themselves may not be linked, so the clustering coefficient could be less than 1. In Fig. 1.1(b), if A and C were linked, they share neighbors B and D, which are unconnected.

NOTES

1. I use the acronym *Internet* to refer to the *architecture* of mobile, digital communication technology as well as the entire *cyberspace* of connectivity. In 1995, Tim Berners-Lee called this the World Wide Web (WWW) or a multi-purpose network of packet switched data.
2. By granular, I mean "small," comparable to grains of sand. In other words, no economic participant is distinguishable in size from another. However, the character of each grain will differ.

3. For more on the Kauffmann Index [1], see Chap. 3.
4. I discuss the term business dynamism in Chap. 3.
5. Joseph Schumpeter writes, "Aristotle no doubt sought for a canon of justice in pricing, and he found it in the 'equivalence' of what a man gives and receives. Since both parties to an act of barter or sale must necessarily gain by it in the sense that they must prefer their economic situations after the act to the economic situations in which they found themselves before the act – or else they would not have any motive to perform it – there can be no equivalence between the *'subjective'* or utility values of the goods exchanged. . . . [However] the just value of a commodity is indeed *'objective'* but only in the sense that no individual can alter it by his own action" [5].
6. Economic models contextualize trade by positing environmental coordinates or exogenous parameters such as number of firms, consumer tastes, production technology, social and political institutions, and organizational structure of trading units. Business strategy is essentially about endogenizing these parameters, as, for example, in changing the organizational architecture of firms from a hierarchical to a flatter layout.
7. While granularity of economic units in the network economy means a decrease in the size of the trading unit, there is concomitantly an *increase* in the number of buyers and sellers when the overall population is fixed.
8. This fairly common phrase is the tag line used by Fareed Zakaria in his weekly CNN broadcast, Global Public Square (GPS). I use it here to provide a perspective on the central issues raised in each chapter.
9. A detailed exposition of graph theory and its application to networks is in the excellent textbook by David Easley and Jon Kleinberg [14].

BIBLIOGRAPHY

[1] Kauffman Index of Startup Activity. Accessed July 3, 2016 from http://www.kauffman.org/~/media/kauffman_org/research%20reports%20and%20covers/2015/05/kauffman_index_startup_activity_national_trends_2015.pdf
[2] Steven Spielberg Commencement Speech, Harvard University, May 2016 (transcript). Accessed July 21, 2016 from https://www.entrepreneur.com/article/276561
[3] Morris, Ian. *Why the West is Winning – For Now*, pp. 190. McMillan, 2014.
[4] Guanzhong, Luo. *The Romance of Three Kingdoms*. Translated by C.H. Brevitt Taylor. E-book distributed by XinXii at www.xinxii.com
[5] Schumpeter, Joseph. *History of Economic Analysis*, pp. 61. Oxford: Oxford University Press, 1954.
[6] McMillan, John. *Reinventing the Bazaar: A Natural History of Markets*. New York: W.W. Norton, 2002.

[7] Dixit, Avinash. *Microeconomics: A Very Short Introduction*. Oxford: Oxford University Press, 2014.

[8] Economist. "The Truly Personal Computer." February 28, 2015.

[9] Smith, Aaron. "U.S. Smartphone Use in 2015." *Pew Research Center*. Accessed April 1, 2015 from www.pewinternet.org/2015/04/01/us-smart phone-use-in-2015/

[10] Rifkin, Jeremy. *The Zero Marginal Cost Society: The Internet of Things, the Collaborative Commons, and the Eclipse of Capitalism*. New York: Palgrave Macmillan, 2014.

[11] Lohr, Steve. "Sizing Up Big Data, Broadening Beyond the Internet", *The New York Times*, August 17, 2014.

[12] Goodwin, Matthew. "Passive Telemetric Monitoring: Novel Methods for Real-World Behavioral Assessment." Chapter 14. In *Handbook of Research Methods for Studying Daily Life*, edited by Matthias Mehl and Tamlin Conner. New York: The Guilford Press, 2012.

[13] Jackson, Matthew. "Networks in the Understanding of Economic Behaviors." *Journal of Economic Perspectives*, vol 28, no 4, Fall (2014).

[14] Easley, David, and Jon Kleinberg. *Networks, Crowds and Markets: Reasoning about a Highly Connected World*. New York: Cambridge University Press, 2010

CHAPTER 2

The Three Drivers: Connectivity, Data and Attention

Digital technology fires on three cylinders to power the network economy: connectivity, the creation and sharing of data, and scarcity of attention.

Abstract DCT has unleashed three powerful drivers: connectivity itself; the collection and parsing of voluminous data and a newly recognized resource bottleneck; and attention. These drivers present opportunities as well as challenges, which enable the network economy to innovate and develop. Connectivity generates transparency as information transfer is more efficient; data enables the creation of patterns and stories about individuals; and the trading of attention (or eyeballs) as a commodity, the central feature of advertising, reveals attention as the new scarce resource. Together, these forces propel a reconfiguration – unbundling and repackaging – of markets and products. Manifestation of DCT across markets is subtle and economic growth is uneven, sticking at the challenges in some sectors, but the race to adapt is enticing and we move forward.

Keywords Connectivity · Data · Attention as a scarce resource · Social capital

Most technology revolutions have addressed fairly practical problems. The steam engine and railroad enabled transportation; electricity enabled

© The Author(s) 2017
S. Bhatt, *How Digital Communication Technology Shapes Markets*,
Palgrave Advances in the Economics of Innovation and Technology,
DOI 10.1007/978-3-319-47250-8_2

production during non-daylight hours. What has the digital revolution done? DCT has enabled us to be virtually connected, and, by unleashing three key intertwined drivers on the economy, has transformed it into a network economy. First, we have instant, continuous, and ubiquitous connectivity, which has created a vast and complex network of connected individuals. Second, this information transfer enables the collection of massive amounts of data. Connectivity and data allow information to flow between market participants, thereby eliminating market intermediaries or traders. In this situation, buyers and sellers directly engage in transactions in a sharing economy where the surplus between the value of the product to the buyer and the cost of production to the seller is shared. Third, connectivity compels us to recognize attention as a scarce resource. The centrality of marketing, of advertisements, for the free consumption of information attests to this scarcity.

Let us consider each of these drivers in turn. What is the implication of the first – that is, of instant, continuous, and ubiquitous connectivity? Connectivity is both real and virtual. Railroads represent real connectivity while digitization generates virtual connectivity, but the underlying technology dynamic has accelerated. Consider that, in the industrial revolution, a whole century separated the 1769 invention of the steam engine by James Watt and the building of the first transcontinental railroad – the Transcontinental Union Pacific – in 1869. Less than a third of that time – only thirty years – passed between the introduction of the Apple II in 1977 and the iPhone (2007).[1] In "The Dynamo and the Computer" Paul David introduces the "delay hypothesis" in a discussion of similar technological time lags: the introduction of electric machinery in the early 1920s took place some four decades after the first electric power station in 1882, two decades separate the discovery of the internal combustion engine and the development of the drive chain that transmitted power to the wheels. The idea is that it takes time for supporting adjustments to be made to the rest of the economic environment before the actual technology has a noticeable impact [15].

One of the most visible results of this new connectivity is "disruptive innovation," or creative destruction, where many businesses from travel agencies and record stores to mapmaking and taxi dispatch have been disrupted. Disruption occurs when newer companies offer cheaper alternatives to products sold by established players and also when existing markets are redefined and the economic landscape reconfigured. Shane Greenstein makes the case that both structural and environmental factors played a role in this process of "innovation from the edges . . . by suppliers who lacked power in the old market structure, who the central firms regarded as peripheral participants in the supply of services, and who perceived economic opportunities outside of the prevailing view" [16]. This economic fluidity extends to new markets with products, heretofore undreamed of, that displace entire industries.

For example, sharing of private goods has been a common feature of society but "sharing" for a price is a novel development. The firm Airbnb involves sharing an underutilized personal space with another person(s) for a fee, crossing the boundary between home and hotel. The hallmark of the network economy is the matching of underutilized resources in market A (to create the supply), with an undersupplied resource in market B (to create the demand). This is often referred to as the sharing economy because the underutilized resource is frequently a privately owned good that is "shared" with others. Another description of the network economy is the "on-demand" economy, which refers to the notion that direct links between buyer and seller create a sense of immediacy in fulfillment of wants.

These direct links result from the elimination of intermediaries, creating a new way of consuming. Connectivity has reshaped the boundaries between markets and firms. Historically, intermediaries had been indispensable for trade to be consummated between individuals due to asymmetries in information, time, and geography. Now we have a TaskRabbit economy where people who want something are instantly connected with those who sell it. Technology has enabled detailed profiles, customer, reviews and rating systems about sellers on social networking sites, which create a compact of trust, reputation, responsibility, and rights (TRR&R), between buyer and seller. Firms have granular data about consumers and can differentiate products to accommodate diverse preferences.

Without intermediaries, the entire surplus, or the gap between value and cost, can be shared by sellers and buyers, with no leakage in commissions. How this surplus is allocated depends upon the bargaining process, and a reasonable outcome to this bargaining "game" depends upon the value of outside options to both parties. What is the buyer (or seller) giving up in order to enter the proposed sharing agreement? However, this is not an entirely rational calculation – emotion plays an important role. Neuroscientists have shown that the limbic system, the part of the brain that is host to attention and memory, is also host to emotion and reasoning. Hence, sellers have to activate emotion in order to capture attention and make a deal. In his acclaimed book, Antonio Damasio elaborates upon this crucial link between reason and emotion:

> ...work from my laboratory has shown that emotion is integral to the processes of reasoning and decision making, for worse and for better...It certainly does not seem true that reason stands to gain from operating without the leverage of emotion. On the contrary, emotion probably assists reasoning, especially when it comes to personal and social matters. [17, pp. 41–42]

The second key driver is vast amounts of static data, commonly called big data (BD), and new ways of acquiring information. Newly created links between individuals generate additional pathways of information gathering. Every connection and transaction, whether it is social, political, or economic, generates information. New links are context dependent – common friends, common interests, and social influence. Patterns emerge from all these nodes interacting and these patterns form the basis for new, dynamic data. These patterns may never be finished, so the network economy is an evolving, complex, and dynamic system and therefore more than simply a static knowledge economy.

BD, therefore, is real-time flow data, not just the stock of past data. It is important to distinguish between raw data and clean data. Clean data is data that has been processed, sorted, analyzed, and conceptualized. It provides information and increases transparency which reduces entry barriers. Transparency forces quick responses from firms, faster innovation, and customization simply to maintain market share, thus empowering the consumer. But it also empowers firms who can tailor their product offering to individual customers with the concomitant price increase.

However, and importantly, on the policy side, the question to be addressed is that of property rights to BD. In the absence of clearly defined property rights, individuals may violate privacy laws as articulated by the 4th Amendment, an issue discussed in detail in Chap. 7. However, there are instances where private data can be valuable public property. For example, in the event of a major health epidemic, vital information about location patterns of infected individuals is more valuable to the governing authority. This data could be accessed from the personal databank of individual smartphones. In the case of city congestion and environmentally sustainable transportation, shared information about traffic patterns and commuting schedules could allow organizations to create smart transportation infrastructure. The aggregation of private information, mostly unsolicited, from participants in the network is crowd sourcing of information, which creates BD. This aggregate body of private information cultivates a diversity of potential solutions to public problems. More generally, crowd sourcing encourages dialogue, develops public understanding of social problems, and motivates action. The example of Jun, Spain in Chap. 5 illustrates this idea.

The key to access any data and avoid privacy infringement issues is to create property rights over personal data so that individuals can voluntarily share their information at the right price. Then, like all personal property or private assets, this will give individuals control over their data. Lessig suggests just such a strategy in "protecting personal data through a property right. As with copyright, a privacy property right would create strong incentives in those who want to use that property to secure the appropriate consent. . . . people value privacy differently" [18]. It is quite possible that a market for this data might lead to exorbitant prices. On the other hand, as is the case today, when data are public property and easily accessible, privacy concerns may lead people to hide data. Like a public park, individuals may not appreciate the full benefits of this shared resource and therefore may not support sharing data or the allocation of resources devoted to its collection. So more thought needs to be given to what the right balance is between making data private property versus public property.

The creation and proliferation of data presents opportunities in two key respects.

The first is recombinant innovation or combinatorial innovation, which involves combining disparate sets of information due to new links. Recombinant innovation is not invention, which is creating something new; it is not improvement, which involves a more efficient way of solving an old problem, much like tinkering along the margin. Innovation is a new way of solving an old problem. This builds organizational capital, which involves new ways of doing business: decision-making, hiring systems, incentive systems, and information flows. Companies "have dispensed with warehouses, trucks and full-time drivers and instead have become middlemen whose sole role is to connect customers with couriers" [19]. Note that this connectivity has eliminated one layer of intermediary along the supply chain – the transportation link. So in effect, the supply chain has shrunk.

The second is the creation of social capital, which consists of shared values and mutual trust. Social capital is created when individuals have repeated trading interactions, inducing a climate of TRR&R, and then form social links in a focal closure. Once formed, social capital generates the opportunity and incentives to create yet more links for transactional purposes (either social, economic, or political). The social ties that bind create membership closure as individuals build trading relationships based on these ties. Social capital enables cooperation in the network economy where economic tensions are resolved via negotiating differences – a point that I will return to later in discussing competition versus cooperation in Chap. 8.

Social capital is bonding capital in tightly connected networks and bridging capital in networks with low embeddedness. For example, the shadow-banking network, which is the unregulated banking network, relies on bonding capital. The network has high embeddedness, with traders having multiple neighbors in common; so mutual trust is the basis of most transactions. Social capital can also exist in networks with low embeddedness, where a strong connection or link between two individuals in two distinct components can foster a bridge, creating bridging capital. For example, in the wholesale diamond industry, social capital is the bridge connecting the wholesale diamond industry in Antwerp, Belgium and Surat, India. In both centers, workers such as diamond cutters, financiers, distributors, and salespeople have long-standing social relationships cementing bonds of trust. Packages of cut diamonds are simply handed over and paid for without inspection. The reputation of each party to this transaction carries sufficient weight so that the diamonds being traded are indisputably adhering to the specifications of the contract.

The third key driver is scarcity of a resource, attention. Monetization of various publishers' digital presence compels them to trade eyeballs on advertising exchanges as they would stocks and bonds. The marketing industry is well aware of this feature. The product underlying real-time advertising exchanges is individual attention. When Google or Facebook places ads on their site, they have sold your attention to advertisers. Search entries on Google and data from Facebook's News Feed, for example, are translated into targeted ads on real-time bidding exchanges.

If attention was private property, with all the concomitant property rights, then individuals could choose to sell their time but, like privately owned land, it would not be appropriated without consent, as done currently by attention-grabbing sidebars and headers on various sites.[2] Encroachment of one's attention due to unsolicited information is equivalent to a violation of property rights. To place this idea into perspective, labor became a form of private property that could be bought and sold for hourly wages only after the Enclosure Movement in the 1500s in Tudor England. Common land was enclosed and transferred to individuals as their private property. Peasants who worked on this land were now displaced and had to trade their labor in the marketplace. Land and labor were traded in a world governed by contracts rather than common custom.[3]

Milgrom and Roberts "interpret 'owning an asset' to mean having the residual rights of control – that is the right to make any decision concerning the asset's use that is not explicitly controlled by law or assigned to another by contract" [21, p. 291]. In this view, unsolicited information packets are encroaching upon private property when they capture attention. Private data, like attention, is subject to similar territorial disputes. Sherry Turkle makes the case that the debate should be rephrased from

> the language of privacy rights to the language of control over one's own data... The companies that collect our data would have responsibilities to protect it... [But] the person who provides the data retains control of how they are used. [22, p. 328]

Note that we have two separate notions of privacy. One is ownership rights over *attention* so individuals have a right to not be addressed, or be left alone. Any information requires attention to be appropriately absorbed. When random bits of information seize attention, they infringe on private property. This could be considered a violation of Fourth Amendment rights to personal property – "The right of the people to be secure in their persons... against unreasonable seizures." The other is ownership rights over personal *data*, which has nothing to do with attention.

MY TAKE

From an individual perspective the critical issue is that of autonomy. Economic agents want to have control over the decision to share attention and data. They want to decide *if* and *how* attention and data are to be used by others. The question of privacy then becomes one of allocation of control and decision-making authority. Subjecting attention to infringement by unsolicited information is equivalent to one's private data being compromised. Both can be thought of as invasions of privacy.

APPENDIX – COOPERATION AND INTERNET ARCHITECTURE

The drivers of the digital revolution – connectivity, data, and time conservation – are themselves born out of the ethos of cooperation. The Internet as we experience it today is only twenty-four years old – it was founded in 1991 by Tim Berners-Lee at CERN. Connecting distributed computing systems, transforming data into the packets of binary code, and routing the data transfer in non-obvious, circuitous paths to conserve time were elements embedded in the history of the Internet. This history is also a story of collaboration and sharing and a prelude to a network economy that is based on cooperation rather than competition. Replacing the linear structure of text with a networked structure is one of the greatest inventions of the digital revolution but these Internet protocols were devised by peer collaboration, and the resulting architecture was one of distributed decision-making and control. It is useful to consider this process of innovation so as to get an overview of the technological landscape.

The Internet is an example of a General Purpose Technology (GPT), the hallmark of which is positive feedback between the technology and associated applications. These co-inventions are generated by customer experimentation and innovation. In this GPT, various computing tasks are assigned to components that are independently or jointly produced. A critical feature of this architecture is modularity, which means that critical components have a limited number of interfaces.

Consider Adam Smith's pin factory and his conception of specialization of labor along different parts of the production process permitting workers to become more efficient at their task and thereby speeding the assembly line. In an important sense there is a coordination problem since workers must specialize along different parts of the assembly line so as to facilitate production. So each worker's choice of specialization depends upon choices made by others. In Adam Smith's world this was made easy by the owner/manager, who assigned work to different individuals. In the world of the Internet, design constraints are standardized by internal trade groups or standard setting organizations (SSO) – the two principle ones are the Internet Engineering Task Force (IETF) and the World Wide Web Consortium (W3C). The former dates to 1969 when the first Request for Comment, a form of collaborative innovation, was published. The International Telecommunications Union is a larger body, currently debating the governance structure of the Internet [23].

Modularity is displayed as layered architecture or a layered protocol stack, which is a sequence of communication and computations commands. The five layers in this architecture are the physical layer consisting of copper or fiber optic cables, the link or router layer, the network and transport layers, and the final application layer. The key feature of this GPT is that it facilitates the process of co-invention via user experimentation and co-discovery. Modularity in this system arises because the five component layers have a limited number of standardized interfaces – sort of like blocks of Lego. The advantage of this system is that as innovation proceeds within a module, it needn't be synchronized with the process in other modules – it can move at a separate speed across modules. Diffusion of an innovation from one module to another is necessarily slower than within modules for this reason.

As a consequence of this asynchronous innovation across modules, economy-wide diffusion of the basic technology is gradual. Apart from the actual hardware, the application layer itself requires supplementary developments in order to be fully implemented. To take an old example, following the discovery of the printing press, the exchange of written ideas could only be consummated after the development of an elaborate network of postal services in Europe by the Tassis brothers of Italy [24]. Similarly, in current times, Amazon has a massive online commerce market but the key link is delivery of these goods and Amazon is reaching beyond UPS, FedEx, and the US Postal service by teaming with Flywheel Software Inc., whose mobile app for taxi service competes with Uber and Lyft [25].

Each layer has its own norms of innovation, so agglomerating the layers for a process of collective governance can be challenging, limiting big push technological changes and allowing legacy systems to survive. This lag between the initial innovation and complementary applications might explain Robert Solow's Productivity Paradox – despite recent advances in digital technology, labor productivity or output per worker in the economy has not increased. (See Chap. 4 for more on productivity.)

Open architecture of the Internet has survived despite the commercialization of content and communication. Neither IBM nor Microsoft, the pioneers, made major parts of the system proprietary. This is a miracle! Why did no firm appropriate the core technology (not the interface at the consumer level) and become a monopoly earning rents or royalties? Copyrights protect an idea, not a product, which is protected by patents lasting for twenty years.[4] However, monopoly status can be conferred to copyrighted software that provides interface protocols and compatibility between itself and other software. For example, Microsoft's copyrights on Windows and Office made it difficult for competitors to overcome the entry barriers into this industry [26].

Currently, there is great debate within the technology community over the value of copyright and the speed of innovation. On the incentives side, there is the argument that copyright confers the rewards of ownership to the copyright holder and hence provides incentives for innovation. Piracy reduces these rewards and could lower innovation. (But see Chap. 6 for more on copyright in the entertainment industry.) On the commons side, the argument is based on the observation that innovation is incremental and builds upon concurrent as well as past innovations and is a collective effort requiring sharing of information. In this case, copyright limits public access to information and hinders creativity and innovation. However, firms could license or cross-license their technology and earn royalties on their innovation, thereby placing it in the public sphere while not giving up revenue rights [18].

NOTES

1. Bill Gates has said, however, that, "the Altair 8800 is the first thing that deserves to be called a personal computer" [15]. The Altair was a machine that hobbyists and hackers, and members of the Homebrew Computer Club in Menlo Park, California, received in a box containing parts that they could solder together and use.

2. Lawrence Lessig makes a more general case that "the protection of privacy would be stronger if people conceived of the right as a property right. People need to take ownership of this right and protect it, and propertizing is the traditional tool we use to identify and enable protection" [18].

3. In medieval times, when each village's economy was isolated, common field agriculture was the custom and institutions were developed such that each laborer had a reasonable land allotment in the common fields. These allotments were scattered and no individual was able to experiment with new ideas or adopt any improvement without general approval but there was also no perceptible social gap between the laborer and farmer. The lord of the manor instituted the process of enclosure, primarily as a means for dispute resolution. Thus originated the institution of private property. The early acts dated to 1773 and were more local than national. However, surrounding a piece of land with hedges and ditches produced "rural depopulation and converted the villager from a peasant with medieval status to an agricultural laborer entirely dependent on a weekly wage." The farmers who owned the enclosed private property benefited due to the increased rents [20].

4. However, due to the 1998 Copyright Term Extension Act passed by Congress, copyrights remain in effect until seventy years after the author's death.

BIBLIOGRAPHY

[15] David, Paul. "The Dynamo and the Computer: An Historical Perspective on the Modern Productivity Paradox." *American Economic Review (Papers and Proceedings)*, 80, no. 2, pp. 355–361 (1990).

[16] Greenstein, Shane. *How the Internet Became Commercial: Innovation, Privatization, and the Birth of a New Network*. Princeton, NJ: Princeton University Press, 2016.

[17] Damasio, Antonio. *The Feeling of What Happens: Body and Emotion in the Making of Consciousness*. New York: Mariner Books, 2000.

[18] Lessig, Lawrence. "Code Version 2.0." Accessed June 26, 2016 from http://codev2.cc/download+remix/Lessig-Codev2.pdf

[19] Miller, Claire. "Delivery Start-Ups are Back Like It's 1999". *The New York Times*, August 19, 2014.

[20] Slater, Gilbert. "The English Peasantry and the Enclosure of Common Fields." PhD thesis, University of London. Retrieved May 26, 2016 from https://books.google.com/books?id=ACEpAAAAYAAJ&printsec=front cover&source=gbs_ge_summary_r&cad=0#v=onepage&q&f=false

[21] Milgrom, Paul, and John Roberts. *Economics, Organization and Management*, 291. Upper Saddle River, NJ: Prentice-Hall, 1992.

[22] Turkle, Sherry. *Reclaiming Conversation: The Power of Talk in a Digital Age*, 238. New York: Penguin Press, 2015.

[23] Isaacson, Walter. *The Innovators: How a Group of Hackers, Geniuses and Geeks Created The Digital Revolution*. New York: Simon and Schuster, 2014.

[24] Mokyr, Joel. *A Culture of Growth: Origins of the Modern Economy*. Princeton, NJ: Princeton University Press, 2014.

[25] Bensinger, Greg. "Amazon Hails Cab for Delivery Test." *Wall Street Journal*, November 6, 2014.

[26] Varian, Hal, Joseph Farrell and Carl Shapiro. *The Economics of Information Technology*. Cambridge: Cambridge University Press, 2011.

CHAPTER 3

The Three Trends: Granularity, Behemoths and Cooperation

Organization restructuring in the network economy is about the unbundling of products and services, both at the production and consumption end, leading to granularity and direct buyer-seller interaction, eliminating intermediaries.

Abstract The first trend, OR, has redefined markets and unbundled products, transforming economic units into granular structures which operate with fewer employees and intermediaries. Connections and data allow buyers and sellers to confront each other directly and transactional integrity is maintained by the pillars of social capital – trust, reputation, responsibility and rights. Agency costs of monitoring and incentivizing workers are lower due to disintermediation. The second trend is an agglomeration of transactions. This trend, enabled by network effects and the trading of attention across multi-sided markets, results in the concentration of economic activity around a few major hubs, the OB. Advertising emerges as a major player in these multi-sided markets. The third trend is a recalibration of the notion of competition.

Keywords Multi-sided markets · Unbundling of consumption · Trust and reputation · US Census data · Economies of scale

In a network economy, economic transactions are characterized by the links between nodes, which represent market participants. The network is

---

© The Author(s) 2017
S. Bhatt, *How Digital Communication Technology Shapes Markets*,
Palgrave Advances in the Economics of Innovation and Technology,
DOI 10.1007/978-3-319-47250-8_3

the market and in a competitive market, prices provide the incentives and information for supply and demand to adapt toward an equilibrium. In this chapter we analyze three developments emanating from connectivity and the vast information resources provided by the Internet: the diminished role played by intermediaries in the network economy; the simultaneous presence of massive organizations or behemoths; and the resolution of this paradoxical outcome by cooperation among agents.

FIRM BOUNDARIES AND AGENCY ISSUES

Firm boundaries are defined by informational and organizational constraints, which limit the span of control of management. A firm can work with outside contractors so long as management can monitor and measure their work. But when opacity of information imposes constraints on monitoring, firms must expand their span of control and envelope the workers into employee status within the firms' boundaries. On the other hand, management will hit the hard wall constraint of limited time and mental bandwidth if the firm boundaries are ever increasing. Synchronizing activities of large numbers of dispersed employees could create an unsolvable logistical puzzle. Therefore, firms will equilibrate at some intermediate size, determined by the industry and the economic environment.[1]

Whenever an agent is interjected between the owner, or principal of an economic enterprise, and its outcome, a problem arises, commonly called the agency problem. In particular, if the nodes with information are not the nodes making the decisions, there are going to be non-optimal transactions. Another way to look at this is if the node, the principal, with the information, resources and authority, delegates decision-making power to his agent, we have a situation where the agent may take a sub-optimal decision because (i) he may not share the same incentives as the principal, who provides the resources and (ii) he may not have the complete information set held by the principal. There will then be conflicts of interest between the two sets of parties, and more importantly, between all the other sets of parties through which the resources and decisions are delegated along the way.

This gulf between management and employees adds complications to the explicit contract which maps worker effort to its outcome. It will, therefore, often need to be supplemented by implicit contracts to overcome monitoring and informational constraints. For example, in addition to wages and work hours, there is an agreement, based on shared expectations, about goals and objectives of the firm. Postulating every eventuality is impossible and hence this shared broader objective incorporates the implicit expectations about worker engagement. Larger firm size increases the complexity of these contracts. The principal may not observe the actions by the agent, yet these actions have enormous implications for the outcome. Sub-optimal production and output decisions (e.g., shirking) can be mitigated by the presence of an intermediary who monitors the actions of the agent to ensure maximum productivity. Disintermediation, as well as production on a smaller scale, can eliminate this agency problem.[2]

There are two separate developments, both powered by digital technology: contraction of firm boundaries and elimination of intermediaries. Shrinking boundaries is evidenced by the fact that small businesses account for over 50 % of all mature firms in the US economy, according to data from the US Census Bureau [28]. Disintermediation is suggested by the changing organizational structure of firms: we are witnessing the restructuring of the production establishment due to the very nature of workers in firms like Airbnb and Uber. Driver-partners, as they are called at Uber, for example, are not employees but independent contractors; so Uber is considered a small firm in employee size. The focus has shifted from hours worked to quality of output. Work done outside the firm's boundaries cannot be monitored and checked for work hours inputted, it can only be judged by output. Hence, the contractual terms between labor and management, the question of control over quality of the product or service and its price, are fundamental to the granular structure of production.

A classic example of this trend toward granularity, and synchronization of principal and agent agendas, is from the music industry. The artist Taylor Swift, as principal, bypassed the intermediating function provided by established record labels and aligned with an independent record label, Big Machine. She had the power to force the world's largest company to change strategy when Apple, as her agent, had to reintroduce royalties in June 2015, for artists whose songs were played during the transition phase as the firm introduced its new app, Apple Music. Apple was rolling out its subscription streaming music service, a free Internet radio station and a platform that allowed artists to upload new songs and videos. During this initial phase, these artists would not earn any royalties, reducing earnings of new artists. This policy was reversed when, in a Tumblr post entitled "To Apple, Love Taylor," Swift concluded with:

> We don't ask you for free iPhones. Please don't ask us to provide you with our music with no compensation. [29]

Swift increased the pressure on Apple by also threatening to withhold her latest album *1989* from Apple's streaming music service.

With no intermediaries, the institutional setting within which buyers and sellers interact becomes increasingly significant. The distinctive feature of this network economy is a cooperative institutional framework, rather than a competitive one. For example, in response to Taylor Swift's post, Eddie Cue, one of Apple's senior executives personally called her to assure the reinstatement of royalties. Importantly, he tweeted an informal message, "We hear you @taylorswift13 and indie artists. Love, Apple" [29]. Cooperation can be interpreted as collaboration or joint-participation so we have market participants collaborating on the terms of transactions.

In order to understand the process whereby intermediaries are eliminated and the supply chain restructured, we need to examine the role played by these market participants. For that we turn to the model of multi-sided markets.

Multi-Sided Markets

Multiple economic agents play a role in any given transaction. Digital coordination of these agents is the role of the platform or online intermediary in this multi-sided markets model. The user or buyer is on one side of the platform, the content provider or seller is on another and the ultimate payee, perhaps the advertiser, is on the third. The platform is a passive intermediary, a bazaar for buyers and sellers, who themselves originate and consummate the transaction. The platform itself requires no deep knowledge of the characteristics or motives of the various economic agents. The platform model captures the essence of a network economy, where the platform now replaces multiple participants along the linear supply chain. This design feature breaks the market into its modular components allowing direct buyer-seller interaction without intermediaries.

By contrast, the traditional economy had active intermediaries who, using historical knowledge of market participants' profiles, systematically matched buyers and sellers. Economists simplify markets by ascribing the matching function to impersonal prices – markets comprise buyers and sellers who respond to prices determined by some invisible mechanism. The price mechanism is insufficient when the interaction consists of "a two-sided matching market that involves searching and wooing on both sides. A market involves matching whenever price isn't the only determinant of who gets what" [30]. Market design is the science of matchmaking, configuring appropriate rules for various markets. The criteria for a successful matching are thickness (multiple agents), lack of congestion (resolving time sensitive issues such as exploding job offers when the offer will vanish after a certain date), and simplicity. The more intensive the care in designing and implementing rules for a given market, the more important are the intermediaries who, in fact, are the actual designers. Al Roth puts it succinctly:

> Not all markets grow like weeds; some, like hot-house orchids, need to be nurtured. And some carefully nurtured marketplaces on the Internet are now among the world's biggest and fastest growing businesses. [30]

Intermediaries, then, are active participants in some matching markets while platforms are passive bazaars.[3] Disintermediation refers to the elimination of active agents intervening in markets, letting buyers and sellers interact directly. Platforms are ubiquitous on the Internet, providing a space for economic agents to transact.

The revenue or business model in multi-sided markets is complex. Consider the media industry, where information content is the product. There are two basic types of revenue models in this industry. Model A where content is free and revenue is based on advertising; model B has the consumer paying for content and there is no advertising; and model C, a combination of A and B, involves both a subscription fee paid by consumers and advertising. Model A is the basis for the Internet radio station, Spotify; Spotify's premium version, however, follows model B. The digital versions of *The New Yorker, Economist* and *The New York Times* are examples of model C, where content is provided by the platform for a fee, but is supplemented by advertising.

The strategic management of all sides of this platform is even more complicated. Consider the transportation and logistics platform, Uber. It has passengers on one side of the market, driver-partners on the other and the Uber platform managing the entire operation. To be successful, Uber needs a critical mass of customers and drivers or it would have the *penguin* or coordination problem. Hungry penguins crowd the edge of an ice flow but fear being the first to dive in, lest there be a predator in the water. Similarly, in platform markets there is a coordination problem arising from the risk of backing the wrong horse. On the other hand, powerful network effects arise if current users recruit new users and the value to users is a function of the size of the network of all users. When more passengers use Uber, more drivers sign up, and it quickly becomes a self-reinforcing cycle. Network effects, defined later in this chapter, can reinforce first-mover advantages so that the initial firm entering the market may have a strategic advantage.

Attaining this critical mass is the major hurdle to success. Do you price low? Do you target gatekeeper nodes? Which side of the multi-sided market should you first address – the customers or drivers/advertisers? Each side fears being stranded, in the *holdup problem*, if the other side (i) becomes too successful and can extract rents, (ii) exits the market, stranding individuals who have developed a dependence on the product or service, (iii) doesn't make an investment in enriching the platform, or (iv) if the platform provider exercises monopoly power by limiting access to certain parties or by adversely impacting the terms of trade. The debate over network neutrality, where Internet Service Providers exercise discretion over content flowing through their distribution network, is an illustration of this latter point.

More generally, in multi-sided markets, should the platform be the center node in a network of suppliers, sales channels and R&D partners? Is the ideal topology a star network, in which other nodes do not have links with each other? Apple and Samsung were both, until recently, examples of a star network, obtaining information from all other nodes and innovating, without the risk that this information is transferred between them [31]. Subsequently, Apple launched its App Store in 2008, allowing app developers to engage directly with one another via the platform. A firm whose network is more embedded can receive help in an emergency or when manufacturing problems arise, but the chance of radical innovation may be lower. Embeddedness creates a more inward-looking network with few links to distinct outside sources of information.

ORGANIZATIONAL RESTRUCTURING – INFORMATION AND DISINTERMEDIATION

With information being instantly, constantly and ubiquitously available, buyers and sellers no longer need the intermediary to perform vital functions characterizing successful markets: matching and secure contract enforcement.

Prior to the Internet age, the central feature of markets were intermediaries who had links with buyers and sellers. When trades are conducted through these intermediaries, not only are direct links important, but also the pattern of links. These patterns determine who trades with whom and at what price. A trader cannot conduct business with an unconnected buyer, who may nevertheless link with sellers, but through other traders. For example, in Fig. 1.1(g), Trader G is connected to only a single buyer, C, but three sellers, and therefore can only trade with buyers B and D via trader A. The reverse also holds: if traders are connected with a single seller and multiple buyers, they can sell only what is purchased from this single seller. The local coffee shop can only sell Starbucks' branded coffee, if it is associated with the chain.

In order to make a profit, a trader must be critical to the trading path. Therefore, if any trader has a critical link to a buyer or seller, such that dissolving this link would eliminate the trade entirely, then that trader would earn positive profits. In Fig. 1.1(g), buyer B can only buy via trader A and seller J can only sell via trader G, allowing both A and G to earn monopoly profits. However, if a seller or buyer is linked to multiple traders, the seller or buyer, respectively, will retain most of the surplus since they can bargain for the best price. This is the case with buyer C who can negotiate a better price due to links with both traders A and G.[4] Disintermediation occurs when a link develops directly between buyer C and seller K, thereby eliminating the trader.

Ordinarily, common interests, common friends and common institutions foster connections between individuals. In the network economy, ease of connectivity itself enables information exchange through these links. For example, a single comment on Yelp could generate a connection between a reviewer of a restaurant and a recent diner, who may disagree with the review and connect directly with the reviewer. This reviewer may be the owner of another restaurant, so this interaction is the key that unlocks connections between these two individuals, who may exist in different parts of the network. The restaurant owner (seller) and diner (buyer) are now linked, with the possibility of trade opening up. This seller may present a product to this buyer who had no previous demand for the good – the new link creates awareness of the possibilities of new consumption or dining experiences and a transaction may be consummated. Buyer and seller are directly connected without an intermediary or restaurant critic.

What is the mechanism at work in this example? It is one that has been in existence since ancient times – word of mouth (WOM). This was an operative feature of most markets until the number of buyers and sellers became uncountable and no common ground existed across all market participants. The intermediary-trader was created to fill just such a structural hole or gulf between individuals wishing to trade.

Virtual WOM has been revived in the network economy through the mechanism of reviews, ratings, user feedback and social media, which provides relevant product and market information, as well as powerful contract enforcement.[5] Information passing along the links of the network economy has not only eliminated intermediaries but also opened up the possibility of new connections, as we saw in the restaurant example above. Recall that we described informative prices as being the defining feature of a competitive economy. Now, the Internet economy, where information is delivered via virtual WOM, is a competitive economy. The information contained in prices is public information, but now, so is private information, which is delivered voluntarily. Voluntary sharing of information via WOM widens the reach of private information.

Differences in preferences, endowments and private information motivate exchange and buyers and sellers need to have trust in their counterparty that contractual commitments will be fulfilled. Reputation, based on crowd-sourced reviews and ratings, is the single most robust approach to creating this trust. In the virtual world, reviews and ratings become powerful WOM enforcement devices due to their ability to become instantly and ubiquitously available. Their effectiveness is due to the fact that with repeat interactions in virtual space, the most egregious punishment is that of ostracism. Avinash Dixit explains:

> Private governance by social groups or industry associations can have advantages of information and expertise, and can use them for arbitration of disputes that the state's courts would find too complex to interpret and adjudicate. These private forums of governance also have at their disposal quite dire punishments for members who violate the norm or code of conduct; they can ostracize the person, or drive him out of business. [33]

And Douglass North traces the origins of self-enforcement to the system of feudalism in the Middle Ages:

> [While] a variety of courts handled commercial disputes, it is the development of enforcement mechanisms by merchants themselves that is significant. Enforceability appears to have had its beginnings in the development

of internal codes of conduct in fraternal orders of guild merchants; those who did not live up to them were threatened with ostracism. [34]

Evidence for Disintermediation

Production Side

OR that has led to granularity and disintermediation is more of a bottom-up process rather than a top-down shedding of corporate divisions. Evidence for this process, discussed below, is provided by the smaller size of new and young firms – startups – which are able to innovate and respond to the drivers of the digital economy. Connectivity has increased information flows to both investors and entrepreneurs so that the information and coordination costs that were a barrier to the flow of early-stage funding have been mitigated. Entrepreneurship is now being viewed as experimentation, which "allows individuals and societies to evaluate businesses and technologies in domains with greater uncertainty than otherwise possible, unlocking deep growth opportunities" [35]. Business dynamism is defined by Hathaway and Litan as "the inherently disruptive, yet productivity-enhancing process of firm and worker churn that reallocates capital and labor to more productive uses" [28]. Business dynamism and entrepreneurship, which were on a secular two-decade decline and attributed to a shortage of capital, have shown renewed vigor as business startup activity has increased [36, 37].

(A) Bloomberg Bata

Using data from Bloomberg over the years 2005–2014 (and Preqin for angel, seed, series and venture debt funding transactions up to $3 billion for private US Internet companies), I find the following trends.

(i) Firms are smaller. Figure 3.1 shows that the percent of US companies that are small companies, defined as firms with fewer than 250 employees, has been increasing by 1.8 % each year over the entire period.

(ii) Employees are fewer. Figure 3.2 shows that the number of employees at small firms has been decreasing at a rate of 0.46 per year or each firm is shrinking by one employee every two years.

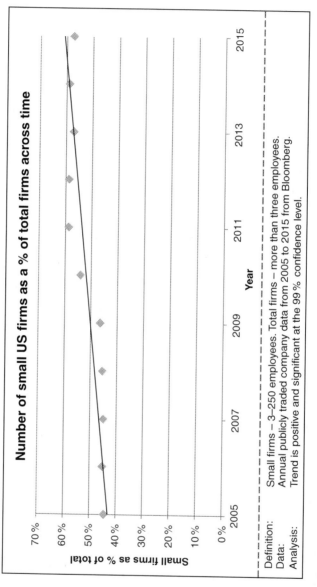

Fig. 3.1   Small firms as a percent of total firms

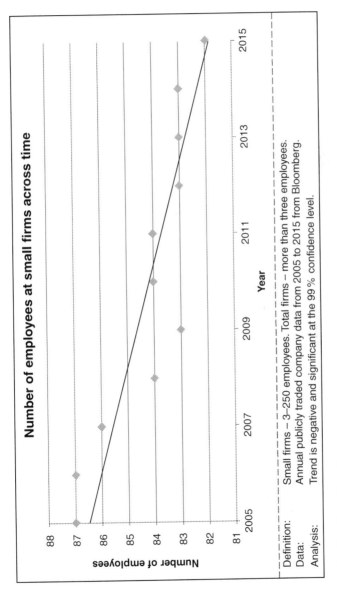

Number of employees at small firms across time

Definition: Small firms – 3–250 employees. Total firms – more than three employees.
Data: Annual publicly traded company data from 2005 to 2015 from Bloomberg.
Analysis: Trend is negative and significant at the 99 % confidence level.

Fig. 3.2 Number of employees

Both Figs. 3.1 and 3.2 support the notion of granularity in firm size.

(iii) Investors are financing very young startups more than mid-stage or later-stage startups. Paralleling the rapid change in technology, seed funding, or very early-stage funding of technology startups has been increasing as shown in Fig. 3.3. The annualized growth rate of seed funding, over the period 2004–2015, is the highest, at a compound annual growth rate of 35.4 %, followed by angel funding at a compound annual growth rate of 10.6 %, and late stage funding growing the slowest, as depicted in Fig. 3.4.

In general, the time line of funding of Internet startups is as follows. At the idea stage entrepreneurs seek seed capital. Angel funding comes next, but is conditional upon reaching certain developmental milestones. Typically, angel refers to an individual, not an investible asset class, but angel investors have been rebranding themselves as seed investors, which might partially account for the phenomenal growth of this asset class. Additionally, startups are now pursuing several rounds of seed funding instead of directly raising venture capital. Finally, there is the staged series funding, each larger than the previous one and each contingent upon pre-assigned growth parameters.

(iv) Initial public offerings (IPOs) in the US technology sector have been increasing at a significant rate. The final stage is the exit stage, where early investors can cash out their investment when the firm goes public in the equity market, in an IPO.

The growth rate of tech IPOs in Fig. 3.5 is lower than the IPO growth rate during the boom years of

the late 1990s and 2000, when hundreds of tech companies went public annually... Today's companies are also waiting longer. In 2014, the typical tech company hitting the markets was 11 years old, compared with a median age of 7 years for tech I.P.O.s since 1980. [38]

One of the reasons that firms can remain private longer is that private investors are fueling their growth, especially in the later stages, as shown in Fig. 3.4.

There is a feedback effect at work in the venture funding industry. The IPOs of companies like Facebook, Twitter and LinkedIn created

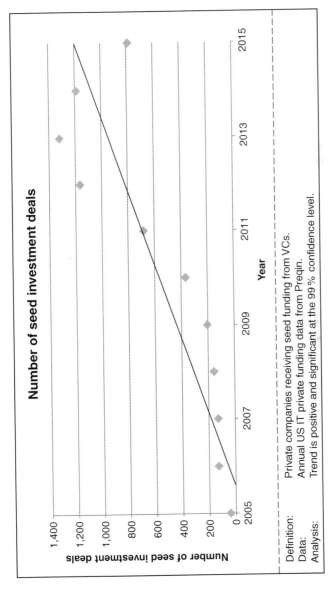

Fig. 3.3 Number of seed investment deals

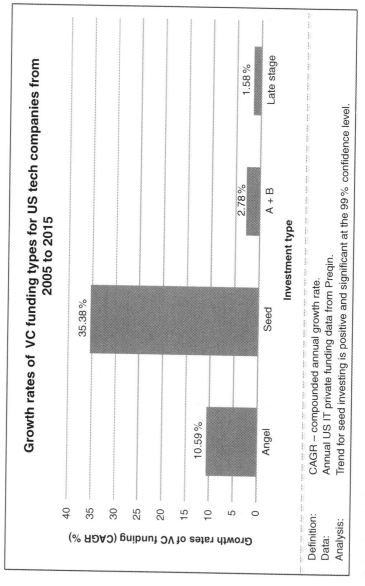

Fig. 3.4   Growth of VC funding

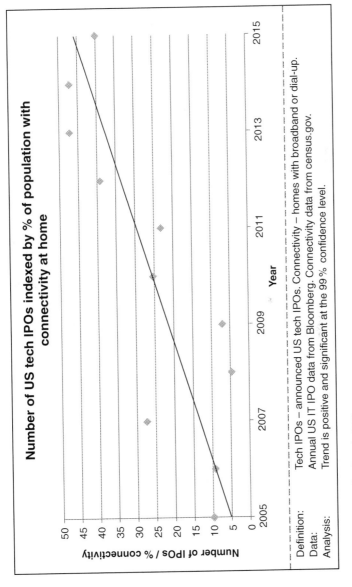

Fig. 3.5 Number of US tech IPOs

significant wealth, both at the individual level and at the fund level, which was then recycled into additional new entrepreneurial ventures. For entrepreneurs, discovering angels was a "kind of trophy collecting" though not without some risks [39]. The growth in early-stage financing of firms, as depicted in Figs. 3.6 and 3.7, supports the observation of renewed entrepreneurial activity and addresses the point made by Dinlersoz et al, who claim that workers in young, entrepreneurial firms have less net worth than workers in the corporate sector and hence, it is financial frictions and borrowing constraints that limit entrepreneurial activity rather than labor market frictions [37].

Granularity or incremental funding in the form of staged investments, as shown in Fig. 3.4, mitigates this financial bottleneck. Initial funding is a bet across a large number of highly uncertain ventures. Most of these bets fail, but the successful and promising candidates are granted additional funds allowing them to scale up. Financing proceeds sequentially and is experimental as each stage reveals more information about the quality of the investment. The funding at each round depends upon the uncertainty that is resolved at that round – will the technology work and does the product have positive sales growth? Choosing to not reinvest is equivalent to not exercising the option to buy further equity investment in the startup.

Contrast this with mutual fund investors who are unable to participate in this staged funding, which requires an early and risky initial investment in order to gain entry into later-stage funding of successful startups. Venture capital firms reverse the process "by starting small and owning a larger share of the firms that turn out to be successful, while attempting to cut their losses on the unsuccessful ones as early as possible" [35].

(B) Kauffmann Foundation Results

Recent data suggest an upward trend in the number of young firms, in addition to an increase in the number of small firms [40].

The Kauffmann Index, using data from the Current Population Survey, was used for the past decade as an important metric of entrepreneurship activity. The new Index focuses on new businesses or startups, and therefore on age of firms, not just size. The Start-Up Activity Index (SAI) rose in 2015 reversing a five-year downward trend beginning in 2010, when it stood at 0.13 declining to –0.04 in 2011, –0.72 in 2012, –0.70 in 2013, –1.06 in 2014, before climbing to –0.37 in 2015 [1].

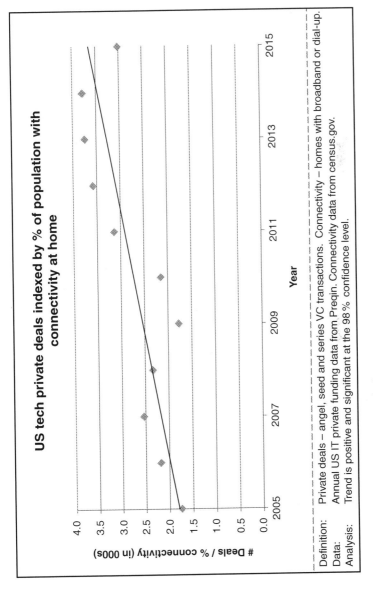

Fig. 3.6 Number of US tech private deals

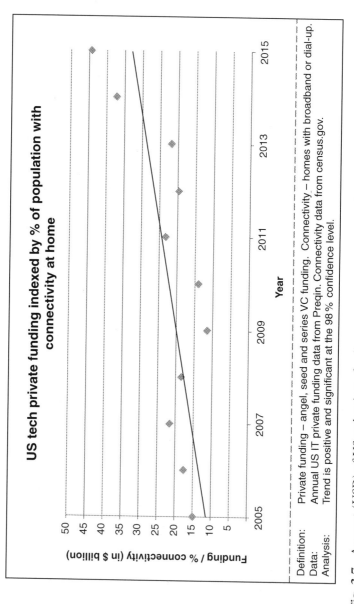

Fig. 3.7 Amount (USD) of US tech private funding

(C) US Census Data

A greater share of employment and a greater fraction of the total number of firms is attributable to *mature* firms (those that have been in existence for more than sixteen years), but a larger fraction of these mature firms is *small* in terms of number of employees. So granularity means small in employee size, although not necessarily young in age.

Aging of the US economy is thought to reduce entrepreneurial activity since fewer new firms are being created and young firms are more likely to fail. Examining long-term trends over the past two decades, Hathaway and Litan find that while there were more mature firms (over sixteen years in age) relative to young firms (one to five years in age) in the US, more of these mature firms were small in terms of having few employees [28]. Importantly, mature, small firms demonstrated the largest growth, as "the growth in firm and employment shares by mature firms has been driven more by smaller firms" [28].

Dinlersoz, Hyatt and Janicki show that the average size (number of employees) of young firms (less than 6 years in existence) has been declining over the years 1983–2012, with the number of employees fluctuating in a band of 7.5 to 9 and the estimated trend showed a significant long-term decline [37].

However, large firms captured most of the *labor* market: they "represent the lion's share of mature firm employment" [28]. The Amazons and Walmarts accounted for less than 2 % of all mature firms in 2011, but as firms were consolidating and multi-establishment firms became more prevalent, their share of total employment was increasing overall [28].[6]

Business consolidation, where large firms account for a significant majority of workers, is different from business network formation. The latter is consistent with the theory of granularity since the nodes of business remain small while there is increasing connectivity across all these nodes. The startup culture can be viewed as a sprouting of small nodes of business, which quickly acquire links to other nodes, some becoming larger hubs. These links provide valuable information about changing consumer preferences, production costs and technology. Business consolidation occurs through acquisitions of younger, smaller firms by larger, mature firms. These acquisitions are often torn apart so as to extract the synergistic parts of the target firms. For example, the engineers are retained, while the CEO becomes a serial entrepreneur.[7]

Consumption Side

Digitization has enabled increasing granularity of products by responding to consumer demand for functionality of products and unbundled product offerings. Production and sales at a granular level lowers the cost of marketing to the appropriate demographic group. Consumers find it cheaper to consume since the initial outlay is minimal and therefore less risky. The sunk cost of making a purchasing error is thereby lowered, enabling a greater sampling of products. The best example of this is the music industry. The music album is unbundled so that music lovers needn't purchase an entire album: they can download, and pay for, individual songs, or they can stream songs. Streaming music is basically a rental service where listeners are paying for temporary access to the music. This allows greater sampling of songs, increasing the market for smaller artists who cannot sustain an entire album. As awareness of smaller bands increases, there may be a concomitant increase in ticket sales and attendance at concerts. For established artists, this effect may not be as pronounced. Apple's new Apple Music service consists of streaming music, a radio station and a platform for artists to upload music and videos. This service provides many smaller pieces.

Amazon, by contrast, started out as a simple online bookstore. Using the information provided by customers using its service, Amazon branched out into ancillary operations that could utilize that information. If I purchased a book on gardening, then it would be useful to connect me to a garden store that sold me seeds and other equipment. Amazon would be the connector or link that brought all my consumption under one umbrella. And so it mushroomed into an enormous company, beyond its startup roots. Can this happen to the other granular companies today? "A small fraction of young firms exhibits very high growth and contributed substantially to job creation," where high growth is defined as firms are increasing their employment by more than 25 % in a year. However, "the evidence shows that most startups fail, and most that do survive do not grow. But among the surviving startups are high-growth firms that contribute disproportionately to job growth" [36]. So Uber could become the next Amazon, but there aren't many such Uberazons.

Recent startups such as TaskRabbit, Airbnb and Uber are examples of small product offerings where the intermediaries and supply chains have shrunk. These firms are software platforms offering unbundled consumption of housekeeping, residential or transportation products. Uber, for example, can be considered a software platform for transport and logistics. Note that, for a company like Uber, the data collected from, and the software created for, daily operations can be leveraged across multiple consumer offerings. Knowing the transportation patterns of frequent users of the car service gives the company detailed information about an individual's lifestyle and possibly the lifestyle of a similar demographic. This information can be utilized or sold for other retail purposes. Such a development gives Uber the opportunity to morph from a startup to a much larger company.

Similarly, in the news media, users can access content via author blogs or via social media sites such as Reddit.com. Facebook's News Feed has a specialized algorithm that feeds you exactly the news that aligns with your profile, so you get to read a customized newspaper. There are also tailored newsletters, delivered via email, such as Ozymandias, which scour the globe for interesting stories that are likely to be of interest to a user's demographic. According to the Pew Research Center, 26 % of Americans get news from a mobile device, 40 % of Internet users favor the ability to customize their news and 75 % of online news consumers get their news from social networking sites or emails [42].

While there is value in getting news stories related to the user's profile, there is noisiness in the information since no editorial supervision is present. While the curator of a blog can actively monitor information, news feeds on social media have no curator. The political implications of this customization feature on news sites are profound – there is limited overlap between individuals with ideological differences. Prior choices reveal preferences and new information flow is sorted based on these choices, thereby reinforcing the insularity and consistency of content. Individuals are grouped into silos and the overall population can become segmented along philosophical and ideological lines, polarizing society. Much of this separation can be attributed to social media, an issue I discuss in Chap. 5.

There is also an unbundling of currency. Whereas public currency, backed by sovereign nations, serves as a payment mechanism and a store of value, private currency unbundles these two functions. Digital currencies or private currencies, such as Bitcoin, serve as a means of payment, but do not serve as a store of value due to their volatility, as discussed in Chap. 6. TRR&R is vital in maintaining a core network of users who will accept this currency as payment and allow it to be sustainable as a standalone currency without the backing of any sovereign government or gold.

ORGANIZATIONAL BEHEMOTHS – HOW DO AMAZON, FACEBOOK, GOOGLE AND APPLE CO-EXIST WITH GRANULARITY?

Google, Facebook, Amazon and Apple are behemoths in a winner-take-all market due to supply side economies of scale and demand side economies. Information has a unique cost structure, unlike most other goods. There are *supply side scale economies* generated by high fixed costs of production and near-zero marginal costs of informing additional people. So it costs zero to sell information to additional users. However, high startup costs will naturally give an advantage to early entrants in the market.

Network effects, both direct and indirect, generate *demand-side economies of scale*, further increasing the early entrant advantage. Direct network effects arise when the value of a product depends upon the size of the network of users as in a telephone network. An important requirement here is implicit coordination among consumers so that all link to the same network. If consumers are spread across different networks, the benefit of network size is dissipated and the direct network effect is lost. The value of joining a social network, say Facebook, increases when more of my friends are also members of this network.

Indirect network effects arise when complimentary goods are produced in proportion with network size as in the case of applications for mobile devices. When more consumers are on the Android network, it pays for more apps to be written for Android devices. Hence, it becomes vital to get to the tipping point or the critical network size in order to collect these network effects. Similarly, when there are more shoppers on Amazon, it pays to be a seller of goods on that site.

Also on the demand side are information cascades. This is the case when behavior conveys information, so if a group of people are making a certain choice then it is may be optimal for subsequent decision makers to ignore private information and follow the crowd. They base their decisions on what others do and ignore their own preferences or private information, leading to a cascade of information. The important feature of information cascades is that people can see what others do but not what they know. Sequential decision-making is also critical to this result as earlier decision makers "share" their information via their behavior. Note that information cascades will increase the network size leading to network effects, which will exacerbate the initial effect. This can result in an outcome where few products enjoy a large number of buyers and many niche products co-exist, an illustration of the Power Law. However, when decisions are made simultaneously and based on information that is independent and identically distributed, information cascades cannot arise because there is no leader to follow – individuals make their choices at the same time.

In the discussion of granularity, we considered the network economy where connectedness mobilized free information flow. However, in some markets ownership of information can be effectively fenced in and managed like a private good. BD is information that can be managed as a private good. Each transaction leaves behind crumbles of personal information, adding to the stock of data owned by the seller. This data set is the resource that sustains behemoths by generating revenue in two ways. First, it promotes efficient, and therefore valuable, advertising by targeting the critical demographic group. Second, it enables superior product development. As more transactions or sales occur, more data are generated, which in turn can be fed into algorithms to target buyers and advertisers. A revenue-generating feedback loop is created and sustained by demand and supply side economies, strengthening the behemoth status.

However, there is granularity at the product level, or product differentiation. Indirect network effects promote links between product sub-groups across these firms, so firms choose compatibility as a strategy to increase their individual market. For example, the iPad and the Kindle Fire are both in the same product category. Having access to a vast library of content via Kindle makes the iPad a more valuable device, which is the indirect network effect. iOS and Android are in the same product category, so that the more smartphone owners that exist, the more valuable is the ownership of smartphones. Amazon Web Services, Google Cloud and iCloud all offer Cloud products to users, encouraging the use of Cloud-based services in the first place. These links across product categories suggest a sort of granularity even while the umbrella organizations are large financially unified firms.

When firms increase compatibility across product categories, they enable users to segment their consumption into functional categories, purchase different categories from separate firms. Functionality becomes the product characteristic, not the business-defined category. For example, the iPad can be a reading device competing with the Kindle Fire, or it can be complementary with the Kindle content library. While the design engineers at the firm choose *compatibility*, the user chooses *functionality*. Take the case of the iPhone. Regardless of the intent of the iPhone designers, the device may be used more as a tool for business information management. Small businesses with dispersed service locations can have their employees send snapshots of their completed work, together with invoices, to central management for more efficient and coordinated billing.

The Amazon phenomenon of winner-take-all can co-exist with the granular market structure. These are markets where a single firm dominates but there is space for a multitude of niche firms that cater to smaller consumer segments, which were previously not served or underserved. The Internet itself can be a driving force of long-tail markets since buyers and sellers meet on the global, digital marketplace. Neither distance nor geography presents barriers to transactions. The term "long-tail" comes from the observation that the distribution of revenues over all firms follows a Power Law. So, for example, the fraction of firms that have k dollars in revenue is distributed as $\frac{1}{k^c}$ where c is some positive number. An important market where this phenomenon is demonstrated is the labor market. There are lots of jobs listing mass-market skills and few jobs for specialized skills. The important point to note is that these specialized skills would have zero listings without the Internet, so the fact that this part of the distribution is "tailing" away toward zero but does not hit zero is the "long tail."

My Take

On balance, how has digital technology revolutionized the organization of markets? I believe that markets will gravitate toward a steady state with both OR and OB. In some markets, commoditization of the product and homogeneity of preferences will support the development of OB. Just as the single village square was central to social life, Facebook could become the global social sphere. Simultaneously, increasing connectivity will enable OR as granularity in products and preferences is unleashed. From independent musicians to chefs, connectivity is likely to lessen the separation between buyer and seller. The structural changes that economists have recently been talking about are in fact a manifestation of this bimodal economy. We have high productivity sectors with OB and low productivity sectors with granularity and OR. There is a shift of labor from the former to the latter, giving rise to lower overall productivity and income inequality [43]. A subtle factor in the productivity debate is the redefinition of labor from employee, with limited control over the terms of the work, to contract worker, who has more control.

Hidden beneath the three trends is another rather stealthy, development accompanying the adoption of digital technology – excess connectivity. If individuals get overwhelmed by the demands of new links being formed continuously and incessantly, there may arise connection fatigue. Each connection brings with it some information which requires attention and processing. Will there be an organic regress from too much connectivity or will there be some external force limiting this? I think we are seeing incipient changes in our social fabric as people sense a loss of self in a sea of multiple identities. It is this sense of dislocation and drifting away from a true identity that will trigger a movement toward solitude and reduced connectivity. Solitude may fuel innovation and reignite productivity in lagging sectors, and business dynamism overall. My own assessment is that we are at a cusp – the connectivity function which was hurtling upwards exponentially, is curving around and slowing down.

NOTES

1. See [20, Chap. 16] for a classic analysis of the determinants of economic organizations: coordination and incentives.
2. As Eric Schmidt and Jared Cohen succinctly write, "On the world stage, the most significant impact of the spread of communication technologies will be the way they help reallocate the concentration of power away from states and institutions and transfer it to individuals" [27].
3. Real estate brokers are intermediaries or active participants in the marketplace for homes, while online sites like Craigslist are platforms. While community moderated, they are free of any active engagement, as signified by their .org domain which symbolizes their non-commercial, public service mission.
4. An analogy to the marriage market is useful. A successful computer engineer based in California on a short holiday to his hometown in India might find it advantageous to hire multiple marriage brokers to increase his chances of returning married. By hiring a single broker, the unfortunate bachelor might be stuck with the one intermediary who has no acceptable female clients.
5. There is some skepticism about user-generated feedback. "There was a lot of fatigue around user-generated rabble that was dominating food dialogue online," Rich Maggiotto, chief executive of Chefs Feed, [a new app], said ... "I am certainly more inclined to trust a chef with a name (and a reputation to guard) than the anonymous grouch who give a great café a low rating after having to wait five minutes for a table" [32].

6. A business establishment is a single physical unit, while a firm can consist of multiple such units.

7. Antonio Martinez [41] gives a fascinating account of tech startups. When his startup, *AdGrok*, was acquired by Twitter in 2011, only the software engineers were part of the acquisition. The CEO, Martinez, joined Facebook, but quit after a few years. It would seem that a startup is then simply a resume, listing the accomplishments of various employees, to be used in capturing the big ticket jobs.

BIBLIOGRAPHY

[27] Schmidt, Eric, and Jared Cohen. *The New Digital Age*. New York: Knopf, 2013.

[28] Hathaway, Ian, and Robert Litan. "The Other Aging of America: The Increasing Dominance of Older Firms." *Economic Studies, Brookings Institutions*, July 2014.

[29] Sisaro, Ben. "With a Tap of Taylor Swift's Fingers." Apple Retreated. Accessed July 3, 2016 from http://www.nytimes.com/2015/06/23/business/media/as-quick-as-a-taylor-swift-tweet-apple-had-to-change-its-tune.html

[30] Roth, Alvin. *Who Gets What – and Why: The New Economics of Matchmaking and Market Design*. New York: Houghton Mifflin, 2015.

[31] Hong, Yoo Soo. "Modes of Combinative Innovation: Case of Samsung Electronics." *Asian Journal of Innovation and Policy*, November 1, 2012.

[32] Bromwich, Jonah. "New Apps to Guide You to Good Food." *The New York Times*, April 10, 2015.

[33] Dixit, Avinash. "Governance Institutions and Economic Activity." *American Economic Review* 99, May 2009.

[34] North, Douglass. "Institutions." *Journal of Economic Perspectives*, Winter, 1991.

[35] Kerr, William, Ramana Nanda, and Matthew Rhodes-Knopf. "Entrepreneurship as Experimentation." *Journal of Economic Perspectives*, Summer 2014.

[36] Decker, Ryan, John Haltiwanger, Ron Jarmin, and Janier Miranda. The Role of Entrepreneurship in US Job Creation and Economic Dynamism. *Journal of Economic Perspectives*, Summer 2014.

[37] Dinlersoz, Emin, Henry Hyatt, and Hubert Janicki. "Who Works for Whom? Worker Sorting in a Model of Entrepreneurship with Heterogeneous Labor Markets." *Center for Economic Studies, US Census Bureau*, CES 15-08, February 2015.

[38] Manjoo, Farhad. "As More Tech Start-Ups Stay Private, So Does the Money." *New York Times*, July 2, 2015.

[39] Isaac, Mike. "For Start-Ups, How Many Angels is Too Many?." *New York Times*, July 7, 2015.

[40] Stangler, Dane. In foreword to "The Kauffmann Index of Startup Activity – State Trends." *Marion Kauffmann Foundation*, 2015.

[41] Martinez, Antonio. *Chaos Monkeys: Obscene Fortune and Random Failure in Silicon Valley*. New York: Harper Collins, 2016.

[42] Purcell, Kristen, Lee Rainie, Amy Mitchell, Tom Rosenstiel and Kenneth Olmstead. "Understanding the Participatory News Consumer." http://www.pewinternet.org/2010/03/01/understanding-the-participatory-news-consumer/. Accessed October 26, 2014.

[43] Rodrik, Dani. "Premature Deindustrialization." NBER Working Paper No. 20935, http://www.nber.org/papers/w20935, February 2015.

CHAPTER 4

The Independent Contractor
and Entrepreneurship in Labor Markets

Jobs are granulized and parceled out in discrete packets to workers across the globe – the common factors are an Internet connection, execution at a distance and output provided on-demand by people who are not employees but independent workers. The online platform for trading goods and services is the on-demand economy, while only services are traded on the online gig economy.

Abstract Empowerment of workers and degree of control are central themes in the workplace. The shift of workers from employee status to independent contractors is accompanied by more control. Employees typically have job security and multiple benefits, but limited freedom over terms of the work. Contractors are at the other end of the spectrum of control, with the ability to tailor their work environment but face the risk of disruptive forces. Jobs are parceled out in discrete packets to workers around the globe, requiring only an Internet connection and execution, while at a distance, is on-demand. As independent contractors, workers own their skill set and, therefore, have incentives, and opportunity, to reconfigure skills using the Internet in online education.

Keywords Young and small firms · Independent workers (IW) · Gig economy · Peer-to-peer (P2P) markets · Productivity

S. Bhatt, *How Digital Communication Technology Shapes Markets,*
Palgrave Advances in the Economics of Innovation and Technology,
DOI 10.1007/978-3-319-47250-8_4

More people in the US work for older and larger firms. The entrepreneurial sector is not driving overall employment in the US. Global economic shocks are better managed by large, multinational firms who have access to advanced DCT, so the decline in entrepreneurship can be attributed to small firms' inability to capitalize on technology [36]. As we saw in Chap. 3, the share of total US employment of young firms, their share of new jobs and the average real earnings of workers in these firms, in nominal terms as well as relative to mature firms, has been falling, dampening business dynamism [37]. The fraction of firms older than sixteen years grew from 1993 to 2011, while the share of firms in all other age groups declined. Further, over the 1978–2011 period, among these older firms, "firms at each size category below 100 employees saw their shares of total employment fall" [28].[1]

The employment picture in US labor markets does not reflect the exuberance in technology-dominated markets. The US unemployment rate, measured as the usually reported civilian unemployment rate, plus all marginally attached workers plus total employed part time for economic reasons, stands at 9.7 % as of June 3, 2016, while the civilian unemployment rate by itself stands at 4.7 %.[2] In the last six years, the ratio of job openings in the information sector to total non-farm job openings has showed a gradual decline from January 2008 (0.0199) to July 2015 (0.0183) [44]. The most recent ratio in June 2016 was 0.0123. More broadly, the expectation that the storm of new devices would generate information-related jobs was unmet: the ratio was 0.038 in June 2007 when the iPhone was released and fell to 0.0199 six months later in January 2008. Similarly, in March 2010 when the iPad was released, the ratio was 0.0188 and rose only slightly to 0.0198 six months later.[3] Employment in the tech sector has not matched the economic possibilities suggested by the introduction of smartphones. Where are all the startups?

These numbers, I believe, misrepresent the underlying fundamentals. The gradual decline in the ratio of job openings in information relative to total job openings from 2008 to 2015 and the fact that the share of young firms and their share of economic activity are declining are not a symptom of slowing innovative economic activity. The structure of the labor market has changed dramatically in recent years with a trend toward increasing granularity in products and services (caused by *mobile* digital technology), a shift in economic power from large institutions to individuals as a proliferation of firms is "owned" by consumers. The latter development has blurred the distinction between supplier and buyer since supplier Y and buyer X in one market may interact as buyer Y and supplier X in another.

These changes are unlikely to be clearly represented in the data, even after accounting for a sectoral shift from manufacturing to services, for four reasons.[4] First, recent legislative action is bringing more flexibility and ownership to potential entrepreneurs, who were previously hampered by financial constraints. Regulation A of the JOBS Act of 2012 reduced financing constraints by enabling firms to raise up to $50 million in equity without being hindered by regulatory approval from the SEC. Second, as discussed in Chap. 3, seed financing, which empowers new firms, is on the rise. Third, labor market frictions in the form of search costs, which had been lower for the corporate sector relative to the entrepreneurial sector, are more likely equalized and lower overall for both sectors due to digitization.[5] And fourth, reduction in transactions costs ignites far-reaching structural changes in labor markets.

The fourth reason is most significant in understanding the labor market implications of granularity. But first we need to describe the kinds of products and services that are amenable to a granular market design and organization. Products that can be fine-tuned or differentiated are more amenable to granularity than the standard commoditized product that scales easily, and therefore is supplied by large corporations. BD is the power behind increased product differentiation. It allows firms to assess consumers at an individual level and correspondingly cater to an infinite variety of tastes, leading to long-tailed markets, as described in Chap. 3. These firms are more likely to be small, in terms of number of employees, making them organizationally nimble in order to adjust to changing and increasingly diverse consumer tastes. Einav and Levin explain that these "peer-to-peer" markets or "market-place businesses" arise due to "variability or diversity in demand, the absence of scale economies in production and of course the existence of well-functioning spot markets to match buyers and sellers effectively" at prices that equalize supply with demand [45]. While digital technology aides in the matching and price discovery functions of markets, powerful reputation effects via user feedback provide the bedrock of trust that enable these P2P markets to function in the first place.

Granularity in product markets translates to granularity in labor markets, as in the Uber model. The Uber model induces underlying structural changes in labor markets. Subsequent to the industrial revolution, work for most individuals was done in relation to a larger institution. These workers, called employees, had a fixed contract that defined hours of work, specified the materials they worked with and provided social security and unemployment insurance. Their contract provided stability in income and work environment while limiting their independence. In the last several decades, this broad social contract and the underlying notions of work associated with it have been dismantled.

Granularity in the supply of labor is manifested in the form of flexibility and ownership. Thus, workers, now called independent contractors, can supply labor in flexible amounts rather than in discrete units.[6] As independent contractors (IC) they "own" their labor or skill-set and can rent it out in discrete amounts. This new labor market development has enlarged the realm of entrepreneurs and small business owners, who formerly were a much smaller slice of the workforce. The gig economy, for example, is characterized by an online platform where people trade services while the on-demand or sharing economy refers to online trade of products and services. Time sensitivity is conveyed by the term "on-demand," which is appropriate since digital connections makes these transactions practically instantaneous.[7]

Workers are now being classified into three possible categories: employees, independent contractors (IC) and independent workers (IW), according to Harris and Krueger [47]. Employee status confers long-term employment but no control over the hours worked or other aspects of the work environment. IC choose when and where to work, and in many instances, also with whom and how. Uber drivers, for example, can choose when and where to work, but not what to charge and whom to drive. The IW category is the new proposed category by Harris and Krueger, who ask whether Uber drivers, who are between clients are "engaged to wait" and should be compensated as IW for that time, or are "waiting to be engaged" and therefore IC. If employees have little control over their work description and IC have complete control, then the IW category is one where the platform "retains some control over the way IW perform their work, such as by setting their fees or fee caps, and they may fire workers by prohibiting them from using their service" [47]. Then IW could have the protections that employees have such as unions, civil rights protections, tax status and insurance pooling.

While there have been flexible work arrangements in the past, the new model makes flexibility more permanent and mainstream. Flexibility does not mean temporary work, performed while looking for another full-time job, working at another permanent job or pursuing another temporary activity such as parenting or studying. Granularity or flexibility is a permanent option! For example, matching recent innovations in ground transportation, an Uber-equivalent for air transport, with smaller operators flying smaller planes to smaller airports is being introduced in the gray charter market. This is the market for airplane ride sharing, in which, business aircraft owners let friends rent their airplanes for flights. "Most US-based charter companies no longer own their fleets. Rather, they manage them for owners looking to generate revenue by chartering out their aircraft – much like time-sharing a vacation house you use only on weekends" [48].[8]

This contract labor market is growing rapidly. Upwork, an online platform for the digital labor market, lists over 2 million businesses with 2500 skills and has over 8 million workers or "freelancers" from 180 countries. They call themselves "digital nomads" and combine work and travel [48]. Some platforms link companies with specific skill sets, such as MBA & Company which links companies with consultants, Topcoder links companies with computer programmers and Upcounsel links companies with lawyers. The long-tail phenomenon is apparent in these markets – Adam Smith's specialization idea has been taken to its logical point in the digital world. Businesses can find special workers to satisfy their particular requirement and specialized workers can find a business that fits their skill-set. Economic data from the Federal Reserve (FRED), show that the employment level of self-employed workers, across all industries, stood at $9.68 million in June 2016, shy of its fifty-year peak of $10.9 million in April 2005 compared with a low of $7.1 million in June 1967 [49]. In past years, self-employment and part-time work was an involuntary choice necessitated by a weak labor market, but today this may no longer be true, as more workers are becoming part of the granular labor market.

There remain two issues that could possibly become roadblocks to the further growth of this contract labor market. One, the IC system might come under regulatory questioning, as suggested by recent developments in the California courts, where former Uber driver-partners are suing the company for lost income. The question is whether the contractor job is considered a temporary "rental" of services or a permanent job with regular employee status. And two, granular markets are faced with the unique problem of allocation of overhead or fixed costs of operation. In typical labor markets, the "firm" that employed labor would bear all fixed costs of operation. These include certification, insurance, financing and taxes. In the case of gray market airplane charters, the operators need to have valid certification from the FAA to maintain insurance coverage on the aircraft and the terms of loan financing. Not only are the owners subject to penalties from the FAA, but the flight crews could have their licenses suspended or revoked.

Harris and Krueger note that basically, workers in this gig economy are at risk from being excluded from a social compact that "represents a synthesis between the desire to enhance the efficiency of the operation of the labor market and to ensure that workers are treated fairly" [47]. There is increased uncertainty in costs and revenues for both the worker and the platform, due to possible litigation or government oversight. Regulatory disparities between labor laws and tax laws exacerbate this uncertainty, and insurance- and unemployment-related protections may be costly since there is no institutional pooling of risk as in company insurance plans.

PRODUCTIVITY AND INCOME INEQUALITY

Two important questions, for workers, arise from the digital revolution. One, how is digitization impacting income inequality? If the talents of individual workers become instantly available on a global basis, will there be a super-star phenomenon where income is concentrated in a few talented individuals? On the other hand, widely disbursed information about jobs in the long tail is opening up opportunities and hence decreasing income inequality. The super-star effect may dominate where the skill set is one-dimensional, like a star basketball player or software engineer. The long-tail equalizing effect may be more important in jobs where "soft" (not easily quantified or measured) skills along multiple dimensions are required. The super-star phenomenon has far-reaching policy implications and circles back to the employment statistics cited at the beginning of this chapter. Shortage of super-star skills relative to the soft skills is what is giving rise to the 1 % phenomenon. Aghion et al. write "there is a whole empirical literature that has flourished over the past twenty-five years which shows that productivity growth is a main component of total growth, that innovation is a key driver of productivity growth, particularly in advanced economies" [50]. They go on to note that "if we look at innovation (measured by the annual flow of patents) and top income inequality (measured by the top 1 % income share) in the US since 1960, we see that these two economic variables follow parallel evolutions." In other words, innovation and inequality move together. Furthermore, innovation also gives rise to increased social mobility as "more innovation implies more creative destruction, i.e. more scope for having new innovators...replace current firm owners." The example of California is cited – it has the highest 1 % income shares and the highest levels of social mobility among all states in the US [50].

The second question concerns worker efficiency. Why hasn't the digital revolution shown up in productivity statistics? The annual percent change in non-farm business sector productivity has been less than 1 % since 2012 (1.09 % in the first quarter of 2012, –0.13 % in 2013, 0.4 % in 2014, 0.71 % in 2015), decreasing to 0.67 % in the first quarter of 2016 [49]. There are three possible reasons.

First, perhaps this is a measurement problem since output is not consistently represented. For example, how does one account for free blogs that provide valuable information to users (content) but are written by unpaid authors? A lot of output now centers on customer experience, as in Apple's ecosystem, which is intangible and doesn't show up in output numbers. A more fundamental measurement issue arises due to the changing organizational structures of firms. This is reflected in the trend toward the "on-demand" economy where P2P markets are operating. Granularity in firms and products leads to many important data points being lost since new organizational forms are slipping under the old radar screen.

Productivity mismeasurement could account for some of this slowdown. Byrne et al. [51] address this question and while they "find considerable evidence of mismeasurement, [they] find no evidence that the biases have gotten worse since the early 2000s." They account for intangible investments; quality-adjusted prices for computer hardware, communications equipment, specialized information-processing equipment and software; globalization; technical innovations in oil and gas production; and a new methodology to account for free digital services such as search and social media. They make the case that consumption of services such as social media, that substitute for other leisure activities, are "properly thought of as a nonmarket production." While these services add value to households, their additional value from advertising-related revenue is small.

Byrne et al. conclude,

thus, there is probably some (at this point very, very small, but likely growing) downward bias in the growth rate of real GDP from the emergence of the sharing economy. It would be useful to have official statistics on the nominal output of the various types of services included in the sharing economy. [51]

It is harder to evaluate changes in overall welfare. There have been value increasing gains to households due to the Internet, but most of these gains are outside the purview of market-sector gross domestic product (GDP) and proposals to include them are debatable. Nevertheless, according to Byrne et al.

> [A]vailable estimates of the welfare gains (based on the value of leisure time) suggests that "free" digital services add the equivalent of perhaps 0.3 % of GDP per year to wellbeing. That is small relative to the 1.75 % slowdown in labor productivity growth in the business sector from 2004–2014.

Second, productivity enhancing innovations are diffusing slowly through the economy, falsely suggesting a decline in economic dynamism – both in the form of fewer startups and slower reallocation of labor resources in response to productivity differentials across sectors [36]. Innovation diffuses more slowly across the broad economy due to the requirement for available complementary technologies. Implementation of a new innovation may take time and hence not show up in productivity statistics. For example, despite the earlier invention of the printing press, a good postal system was crucial for spread of printed matter across most of Europe. Not until the Tassis brothers organized the courier system much later in the fifteenth century, linking important cities in the Holy Roman Empire, did the impact of the printed technology materialize [39]. Today, users may need to figure out the various productivity enhancing aspects of their devices, like the iPad, beyond the stated function. The functionality of new products and services is often developed by users, who customize the product's functions to their own requirements. While the iPad was introduced in 2010, its present day use in restaurants and pre-schools is an unforeseen and, perhaps, fortuitous development.

This slow diffusion of technology shows up in the weak productivity statistics. Structural changes in the economy precipitated by shifts in the labor market from more productive to less productive sectors can give rise to a slower growth rate of productivity despite technological advances in many industries. Dani Rodrik says "This perverse outcome becomes possible when there is severe technological dualism in the economy and the more productive activities do not expand rapidly enough." This "growth reducing structural change has been happening recently in the United States" [52]. For example, the Food and Beverage industry's share of full-time equivalent workers grew by 0.22 % over 2010–2014 but declined in productivity over the same time period, while Broadcasting and Telecommunications (B&T) saw its employment share shrink by 0.16 % but its productivity grew [53]. In fact, productivity growth weighted by the industry's share of full-time equivalent workers is highest for B&T among all industries. Disconnecting ownership from consumption in the B&T industry, an issue I discuss in Chap. 5, may have enabled more rapid diffusion of DCT in the broader communication, information, media and entertainment (CIME) industries.

Skill Enhancement and Digitization of Learning

Both the employment picture and the weak productivity spillover lead one to question the role of technology in higher education and primary education. Do workers not have the requisite skills to compete in the network economy? Can we harness digital technology to unbundle the teaching process, splitting knowledge into two skill groups?

One group, consisting of complex, but repetitive, skills can be moved to online courses, such as MOOCs. The second group, non-repetitive skills, requires an interactive component and, therefore, can only be taught in a lecture- and discussion-based format in a brick-and-mortar environment.[9] The moments of actual learning happen when students are engaged and empowered in interactive discussions. So, perhaps, the role of professors is that of a guide or coach on the sidelines, watching players and correcting flaws in reasoning. The Socratic teaching method is an example, where material is taught and learnt through asking and answering questions rather than one-way lectures.

An important feature of interactive discussions is the network effect. When more students are participating in a discussion, the value of learning increases in proportion to the size of the group. Clearly, this benefit reaches a maximum at some critical group size, since too large a group leads to chaotic learning. According to William Bowen, "a great advantage of residential institutions is that genuine learning occurs more or less continually, and as often, or more often, *out* of the classroom as in it." [54]

The evidence on the usefulness of MOOCs is encouraging. In a Carnegie-Mellon University study on a hybrid-mode statistics course, William Bowen and his colleagues found [first,] no statistically significant differences in standard measure of learning outcomes (pass or completion rates, scores on common final exam questions, and results of a national test of statistical literacy) between students in traditional classes and students in hybrid-online format classes. Second, this finding is consistent not only across campuses, but also across subgroups of what was a very diverse student population. [54]

The upshot is that while measures of learning did not substantially change, widely divergent student groups benefited from this hybrid format when compared to the traditional brick-and-mortar only format. What is remarkable about this result is that the worry that learning outcomes will be compromised with the online hybrid format is without basis. This may have important implications for cost-containment in higher education. The conclusion is that continued incremental approaches to this mixed learning format might generate even more favorable outcomes. Along the lines of granularity in the labor market, Bowen says that "MOOCs could allow students to *experiment* with different classes and get a sense of what disciplines interest them before they ever set foot on campus."

Artificial intelligence researchers have made the important point (Moravec's Paradox) that complex reasoning, which requires precision and regularity, is hard for humans but easy for machines, but tasks that require generalization, perception, creativity and interacting with the real world (low-level sensory motor skills) are relatively easy for humans but computationally expensive. Any task that can be described by an algorithm or is repetitive, then, will be outsourced to technology. Pattern recognition, recombinant innovation, multi-sensory communication and developing creative solutions to previously unimagined problems are traits that will be in high demand since they cannot be converted to an algorithm. These are the skills learned in an interactive environment, and are particularly well suited for young adults born between 1980 and 2000 (the millennial generation) who are accustomed to a world, ushered in by the introduction of the iPhone in 2007, of interaction and transparency. The connections provided by learning in a communal setting are fundamental to acquiring non-algorithmic skills. By unbundling basic learning from the interactive component, MOOCs might present a platform for increased interactive learning and foster innovation. Steve Pinker writes:

The main lesson of 35 years of AI research is that hard problems are easy and easy problems are hard. The mental abilities of a four-year-old that we take for granted – recognizing a face, lifting a pencil, walking across a room, answering a question – in fact solve some of the hardest engineering problems ever conceived. [56]

Social media pervade the lives of millennials, who

work more closely together, leverage right- and left-brain skills, ask the right questions, learn faster and take risks previous generations resisted. They truly want to change the world and will use technology to do so. [57]

Organizational structures are adapting to this demand for transparency and inclusiveness which are critical for successful innovation and learning. As firms become granular and each worker is empowered with more responsibility, they will require an open culture of sharing information in order to perform their own task efficiently. There is a natural inclusiveness when workers, particularly in newly formed organizations, participate in formulating the mission and strategizing the goals of their firm.

Companies are now using social networks to create a pool of willing and competent candidates in the inventory of prospective employees. Maintaining ties with recent job applicants and former employees is valuable in screening candidates. The older hiring process requires intermediaries, who post a job, receive resumes and conduct interviews, all on behalf of clients. The interactive environment made possible by social networks facilitates deeper skill-integration of employer and employee. For example, last year Zappos planned to have candidates join a social network called Zappos Insiders where "they will network with current employees and demonstrate their passion for the company" in the expectation that the company will call them should the need arise [58].

My Take

In labor markets, workers are demanding the same kind of autonomy that we discussed in the case of the market for individuals' attention. Labor is not merely a single factor of production but also owns the means of production, or capital. Specialized knowledge of a task will empower the worker to manage and control the way work is done. The clean delineation between workers, management and owners of capital is fractured as the former are also owners of human capital. Skills that cannot be automated are an increasingly important resource in the digital economy, so acquisition of these skills will generate powerful human capital. An alignment of objectives will be pervasive as owners of financial capital are also owners of human capital, leading to a flatter organizational architecture. No more bosses!

NOTES

1. Note that while young firms (one to five years in existence) are generally small (less than 250 employees), not all small firms are young.
2. Marginally attached workers are those who are no longer looking for work, for whatever reason.
3. Author calculation based on data from the Federal Bank of St. Louis (FRED) database. By contrast, in October 2001 when the iPod was released (actual date was October 23, 2001), the ratio was 0.018 and was 0.024 six months later.
4. In addition, the trend toward granularity is not reflected in the papers cited above since their data series ends in 2011.
5. Dinlersoz et al. [37] segment the labor market into the entrepreneurial and corporate sector, which are associated with young and mature firms, respectively.
6. I will call all labor by the generic name, workers, which could be office workers, manufacturing workers, agricultural workers, etc. The Internal Revenue Service treats employees differently from contract workers since the former file W-2 forms versus the latter fill out 1099 forms.
7. These terms gained global publicity after Democratic Presidential candidate, Hillary Clinton's address at The New School in New York, in July 2015. She said, "This on-demand or so called 'gig' economy is creating exciting opportunities and unleashing innovation, but it's also raising hard questions about workplace protections and what a good job will look like in the future" [46].
8. Additional examples from the transportation industry are UberChopper (for helicopters) and Sailo (Uber for boats). Many thanks to Pia Sur for suggesting these examples.
9. Jeffrey Young writes, "But there's a growing sense that monologues by professors are of limited effectiveness for many of today's students. The teaching style is a tradition passed down through generations of academics, and despite the addition of computers, projectors, and PowerPoint, little has changed in the basic model: A professor talks, large numbers of students listen, and one or two brave souls ask questions in the final moments. Class dismissed" [55].

BIBLIOGRAPHY

[44] US. Bureau of Labor Statistics. *Job Openings: Information* [JTU5100JOL]. Retrieved from FRED, Federal Reserve Bank of St. Louis https://research. stlouisfed.org/frcd2/series/JTU5100JOL/, September 5, 2015.

[45] Einav, Liran, Chiara Farronato, and Jonathan Levin, Peer-to-Peer Markets. (August 2015). NBER Working Paper No. w21496. Available at SSRN: http://ssrn.com/abstract=2649785

[46] Lapowsky, Issie. "Hillary Clinton will Crack Down on the Contractor Economy." Wired Business, July 13, 2015. Accessed July 3, 2016 from http://www.wired.com/2015/07/hillary-clinton-gig-economy/

[47] Harris, Seth, and Alan Krueger. "A Proposal for Modernizing Labor Laws for Twenty-First-Century Work: The Independent Worker." *The Hamilton Project*, Discussion Paper 2015-10, December 2015.

[48] "Gray Market Charter." Business Aviation Insights, January 30, 2012.

[49] Upwork website. "Get More Done with Freelancers." https://www.odesk.com/?vt_cmp=US_oDesk%20Brand&vt_adg=oDesk&vt_src=google&vt_med=text&vt_device=c&vt_kw=odesk&gcl. Accessed November 7, 2014.

[50] Aghion, Philippe. "Comment on 'So What Is *Capital in the Twenty-First Century*? Some Notes on Piketty's Book' (by János Kornai)." *Capitalism and Society* 11, no. 1 (2016). Article 4.

[51] Byrne, David, John Fernald, and Marshall Reinsdorf. "Does the United States have a Productivity Slowdown or a Measurement Problem?." *Brookings Papers on Economic Activity*, March 2016.

[52] Rodrik, Dani. "Innovation is Not Enough." *Project Syndicate*. Retrieved on June 9, 2016 from https://www.project-syndicate.org/commentary/innovation-impact-on-productivity-by-dani-rodrik-2016-06?utm_source=project-syndicate.org&utm_medium=email&utm_campaign=authnote

[53] Vollrath, Dietrich. "More on Decomposing US Productivity Growth." Retrieved on June 9, 2016 from https://growthecon.com/blog/More-Decomp/

[54] Bowen, William. *Higher Education in the Digital Age*. Princeton, NJ: Princeton University Press, 2013.

[55] Young, Jeffrey. *Beyond the MOOC Hype: A Guide to Higher Education's High-Tech Disruption*. Amazon Digital Services, Inc., and excerpted in *The Chronicle of Higher Education*, November 7, 2013.

[56] Pinker, Steven. *The Language Instinct: How the Mind Creates Language*. New York: Harper Collins, 2010.

[57] Agan, Tom. "Embracing the Millennials' Mindset at Work." *New York Times*, November 9, 2013.

[58] Auriemma, Adam. Zappos Zaps its Job Postings. *Wall Street Journal*, May 26, 2014.

CHAPTER 5

The On-Demand Economy and How We Live: Communication, Information, Media and Entertainment

Access to media content can be achieved by temporary ownership or renting, and not purchase, and by consuming fractional portions rather than the entire package. This creates demand for novelty as it shortens the life cycle of any given message.

Abstract DCT has provided choices over content and the vehicle for consuming media content. Unbundling enables the authors of content, the sellers, to be separate from the distributors, or buyers, of their creations. This induces both sides to innovate in the creation, curation and dissemination process, as exemplified by user-generated content in social media and product reviews. Consumption of content is a choice between ownership of individual units in the content bundle; renting via streaming or temporary ownership and the traditional format of purchasing the entire bundle via download. The distinction, from a consumption point of view, between a product that is owned and a service that incurs a one-time fee has been dissolved – access is all that matters.

Keywords Ownership versus access to content · Creation · distribution and delivery of content · Social media

S. Bhatt, *How Digital Communication Technology Shapes Markets*, Palgrave Advances in the Economics of Innovation and Technology, DOI 10.1007/978-3-319-47250-8_5

Granularity in the CIME industries is manifested by multiple consumption formats: ownership of individual units in the content bundle; renting or temporary ownership of units in the content bundle; and the traditional format of owning the entire bundle. The distinction, from a consumption point of view, between a product that is owned and a service that incurs a one-time fee has been dissolved. Access can be obtained by purchase or rental agreements, making ownership a redundant contractual arrangement.

In this chapter, I first address choices facing buyers and sellers of media content and then explore the nature of the content. An important form of crowd sourced content, social media, is discussed in the second section, in the context of engagement, empowerment and immediacy. The business model of the CIME industries sits on a core revenue foundation provided by the advertising industry, examined in the final section of this chapter.

THE MARKET

Streaming music is the latest development to change the landscape of the music industry. The streaming revolution is proving to be more profound than the move to digital, as revenues nearly doubled over 2014, growing by 45.2 % [59]. Instead of buying an entire album, users can rent songs or stream music by subscribing to a service such as Spotify.

Record companies have adapted their business model from ownership of music to access to music, which is reflected in the growing share of subscription and streaming revenues as a percentage of global, digital revenues. The crucial fact is that global music revenue growth (3.2 %) was positive, after nearly twenty years of decline. Consumers have more choice in music offerings, and streaming is the high growth segment; subscription services grew by 58.9 % while pure ad-supported streaming services grew by 11.3 %. The industry derived 45 % of its global revenues from digital services exceeding revenues of 39 % from physical format sales, with 14 % from performance rights and 2 % from synchronization revenues in 2015 [60].[1] Crucially, however, artists, and the music community, are concerned about the "value gap" where they are not being fairly compensated for their content as original rights holders. The Taylor Swift discussion in Chap. 4 illustrates exactly this issue.

Customized product differentiation further distinguishes the product offering: for instance, while in the past Spotify sorted music solely based on genre, it now offers music associated with an activity such as relaxation, meditation or exercise. This allows listeners to cross boundaries and rapidly discover new artists, adding to the song's ranking since the music industry's ranking index, the Billboard charts, now count plays on Spotify and YouTube in their calculation of the country's top hits. There is a strong feedback effect here since programmers for radio stations use these charts to create their music rotation list. From the perspective of the content creator or artist, the choice offered by streaming music services increases competition so many of the new, undiscovered artists can become one-hit wonders.[2]

In addition to granularity, the intermediary or the producer is eliminated as in the new model recently espoused by the band Radiohead on BitTorrent.[3] Prior to digitization, it was not financially viable for artists to release their own album. With the virality offered by the digital music model, promotion and distribution of music is essentially free. Internet radio and apps such as Spotify are examples of avenues that an artist could use to promote music. In some cases, the artist only has to incur the cost of recording and producing the music; most of the marketing has already been done by the record label. Radiohead is an established band, performing since 1992. The full album by Radiohead can be downloaded only after the consumer pays – BitTorrent keeps 10 % of the fee and the rest goes to the artist. In this business model, the creator of information gets to keep most of the revenue. BitTorrent requires you to use the same computer and IP address to re-download the album at a future point. By creating innovative distribution deals with artists, the company is overcoming the free file sharing issue.

Music upload services, such as YouTube, are exempt from many licensing fees due to early music industry legislation that was intended to protect passive listener intermediaries from copyright liability. For example, user-generated content on YouTube might have a trending song playing at some public event. Is this copyright infringement? YouTube has automated software that evaluates the situation and requests the user to remove the material in the event of copyright violation. But there is no personalized monitoring, either by the artist or by the digital platforms.

The central issue in CIME industries is copyright protection. The standard argument is that weak copyright protection shrinks revenues for the producer of original content and, more importantly, reduces the incentive to invest in creative content since the owner may not be able to recover the high initial investment. Revenues for both the artist and producers stem from subscriptions, advertising, legal digital downloads, physical sales and live concerts.

Weak copyright protection can, paradoxically, increase revenues. In the music business, after Napster made large-scale file sharing (mp3) possible since 1999, there was an increase in the supply of concerts, as artists shifted to live performances, and also an increase in demand for live performances, since they are perceived as complementary to digital music. File sharing increases demand for albums due to the sampling effect as information about new music is now easily available. Traditional radio and Billboard top 200 lists used to be the sole source of information, but are now replaced by Internet radio and commercial uses of new songs or synchronization deals where songs are played in advertisements, films and television programs such as major sporting events. Sampling of individual songs creates more choice and enables the user to purchase a wider variety of music. Network effects, where users value music in proportion to its sales rank, will spread this popularity, leading to even more album sales. For small bands, concerts ticket sales and album sales both increase as digital technology increases awareness of these bands [62].

The unauthorized digitization of information goods and services, like music, can reduce the number and price of the original works supplied. But piracy in the music industry due to P2P file sharing via BitTorrent has an interesting feature. Downloading speed for music depends upon the number of initial seeds or legal sales of the music for the obvious reason that locating the music of a little known artist will be much more time consuming than that of a well-known artist. Once a song has been legally downloaded, the download speeds for subsequent users increases, which leads to more downloads and hence increases visibility of the artist. While there is an incentive, both for the artist and the producer, to incentivize more legal sales to start the seeding process, further downloads are enhanced by piracy, as direct network effects strengthen this process leading to possibly higher revenue overall. Consequently, extracting rent or higher price from initial, high value customers could offset the lost revenue for the producer from subsequent, late stage piracy. So the net effect of piracy may actually be increased revenue both for the artist and the label [62].

There are two aspects of copyright. First, there is the question of unintended use of copyrighted material. Second, there is free advertising for the artist. If there is to be compensation in the first case, then perhaps the publicity should also be accounted for. It was clear when Napster was taken down that companies that actively distribute music online were going to face issues of liability for hosting content. Artists and copyright holders would prefer licensing fees at every point along the value chain but monitoring the distribution of content in real time was challenging. The negotiated compromise was the "safe harbor" agreement, as part of the Digital Millennium Copyright Act (DMCA), which stipulated that hosting firms were not liable for copyright violations until they were notified.

Customization

The ability to extract data from streaming demand has enhanced the marketing ability of record labels and platforms, who can more effectively tailor their products to user demand. They have more instantaneous data with knowledge of when and where tracks are played and stored, whom they are shared with, and the quality and depth of user engagement. When a song is featured on streaming playlists, companies can actively promote that track on radio which can then spur demand for downloads to personal playlists. For example, OMI's *Cheerleader* became a hit after Sony Music examined streaming data and subsequently promoted the track [59].

On the demand side, changes in search technologies allow consumers to find a wider variety of music, movies and books making the long tail more accessible. Recommendation or personalization engines enhance product discovery and produce long-tail sales distributions as niche products are revealed. Moreover, consumers can now view samples of book pages, listen to sample music tracks, and watch scenes from movies on multiple devices. The information channels have multiplied beyond radio publicity [62]).

While media platforms have the power to choose which artists and which content to promote over their channel, established artists have the power to choose over which channel, among multiple platforms, to distribute their music. They can sell albums on CD or via downloads or by streaming their new songs immediately. In the latter case they risk losing album sales, but they gain publicity. Artists like Adele, Beyonce and Taylor Swift "have accumulated a rare level of power in the industry, allowing them to take risks over how their music is released and consumed, and the rest of the business has taken notice" [60].[4]

The Seller – Unbundling the Medium and the Message

In the network economy, granularity allows unbundling of the medium and the message in the media industry. Delivery of content, distribution of content and creation of content are three separate services. The medium or physical delivery of content is performed by underground cable, satellite or radio waves. Distribution of content is managed by broadcast television networks and platforms that distribute over the Internet, called OTT (over-the-top). Creation of content is by broadcast television networks, HBO, Netflix and many other independent producers.

In the traditional model, large organizations packaged the message and the medium. Broadcast television and the major movie studios, such as Warner Brothers, Universal Pictures and Sony Entertainment, generated the content and controlled the distribution. MGM was even more vertically integrated: it produced movies, marketed or distributed movies and had ownership interests in the delivery via the theatre chain, Regal Entertainment Group. The *New York Times* hired the journalists, marketed the paper and printed the content. AT&T marketed the telephone devices and owned the copper wires, via its wholly owned subsidiary Western Electric. Sony Entertainment sold albums as well as hired and marketed artists. The sellers effectively bundled the final product along with necessary complementary goods.

In the network economy, the entertainment behemoths, whether they are movie studios or broadcast television, no longer control the flow of content to users who can access similar or identical content via the Internet. Figure 5.1 explains the fragmentation in the industry. Content creators and users flank the central delivery channels, with content on the right and users on the left. Riding astride the central delivery channels are the distributors or OTT platforms. This depiction of the industry makes the unbundling process clear – whereas in the pre-Internet days there were only two players in this diagram, the users and the producers, today we have four: users, creators, OTT distributors and delivery service providers.

Who are these players? Content producers such as app developers, advertisers and Netflix, are creators of the Internet value chain (the right segment in Fig. 5.1) and consumers are the ultimate users at the end of the chain (the left segment in Fig. 5.1). Internet service providers (ISPs) provide the physical infrastructure or the bottom layer of the intermediary and offer the last mile connection or access to the consumer network.[5] On top of these ISPs are the OTT distribution platforms such as Facebook, eBay and Google, who act as intermediaries between users and content producers.

There is further unbundling at the distribution level as product composition, pricing and revenue sharing are untangled. Digitization of media, with both downloadable and streaming content, enables multiple distribution channels (the medium), which are not integrated with

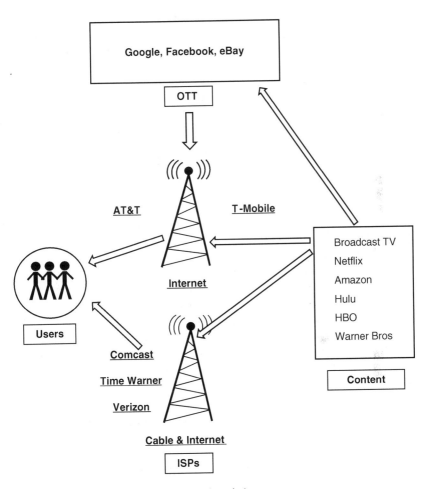

Fig. 5.1 The digital entertainment value chain

producers of this content (the message). Downloaded content is purchased content while streaming is the rental platform. Users can access content via one or both of these formats or multi-home, which suggests that these alternate consumption channels are not substitutes but rather complementary products. Downloadable audio content or Podcasts, previously limited to radio broadcasts, is available on Audible by Amazon, iTunes by Apple and Audioboom in the UK. Downloadable video can be consumed as short clips on YouTube. The rental platform, or streaming video, pioneered by Netflix, is offered by Amazon and Hulu, and streaming audio is offered by Pandora, Spotify, Apple Music, Dweezer, YouTube, Tidal and others.

Multiple distribution channels grant content creators the power to make their product available either exclusively or sequentially on multiple platforms. There is no single digital jukebox. Music platforms such as Spotify, Apple, Pandora, iTunes, SoundCloud, YouTube compete, as artists strategically withhold their music from different apps. For example, in March 2016, the rapper Jay Z removed some of his albums from Spotify and Apple in favor of a streaming service, Tidal, that he owns jointly with Kanye West, who released his latest album *The Life of Pablo* only on Tidal, making it unavailable for sale. Beyonce, Rihanna and Coldplay, all partners in Tidal, also released music favoring that service. Meanwhile, Adele bypassed streaming her newest album *25* and Taylor Swift removed all her music from Spotify in November 2014, criticizing the business model, which made music freely available. After much public debate, Adele released her only her acclaimed single, "Hello" for streaming and purchase on October 23, 2015; the entire album was released for purchase, not streaming, on November 20, 2015, by Columbia Records, a label owned by Sony Entertainment. If the two channels are complementary, streaming music should increase visibility and demand for the album. In the US the album sold 3.38 million copies in its first week after release, beating the single-week maximum for an album since 1991 [60]. This strong correlation is suggestive of the advertising power of streaming song tracks prior to actual album release.

When consumers access content via multiple channels, news media face the key issue of multi-homing. They must address not only simple issues pertaining to content, but also the intermediary through which it is being accessed. Currently, the *New York Times* has a digital version, an application for mobile devices and a print version, all offering slightly different content as well as different advertisements. Creating a following on Twitter or Facebook increases traffic to an established media player's digital or print version, further increasing the reach of the news, but not without its drawbacks. The immediacy created by multi-homing generates side effects. When some event occurs, and before it is fairly reported with accurate facts, social media elicit quick reactions from a general user base and often "the knowable, verifiable truth is left in the dust . . . why slow down and wait for clarity when there's an angle to promote, a grievance to air. Damn the torpedoes and full screed ahead," according to *New York Times* columnist Frank Bruni [63].

The Buyer – Unbundling Ownership and Access

Unbundling allows a distinction to be made between ownership of content and access to content, a possibility not recognized by older, established firms. Rather than pay a single price for the bundled good, users can now buy or rent individual components, priced as a single upfront price for downloads or as a recurring rental fee for subscription streaming services. Purchase enables users to access content for an unlimited time horizon while the rent format, paid for by a subscription fee, allows access for the duration of the contract.

This view enables the music industry in general, and the artist, in particular, to price their content by incorporating user discount rates. Users exhibit time sensitivity when they demand the four As or All the music, Anytime, Anywhere and on Any device. In this context, discount rates are a metric of user time-sensitivity. When the user is more time-sensitive, the discount rate is higher.[6] In this case, the urgency for consumption is higher so ownership is preferable, unless subscription fees on streaming content are sufficiently low. High-speed broadband is the key bottleneck since delivery of content remains the province of the Internet service provider, which is often combined with cable service in many parts of the country. The introduction of gigabit fiber networks (speeds exceeding one gigabit per second) by Google and others is likely to lead to a large increase in average download speed offered by cable providers, making streaming content easier to access [64].

Content and the Nascent Behemoths

The boundaries between the world of broadcast media, which include CBS, NBC and HBO, and the world of the Internet, with firms like Netflix, Hulu and Amazon, are colliding [61]. As consumers watch content directly on the Internet, the market for entertainment is redefined from large, vertically integrated companies to well-curated libraries of content, such as Amazon, Netflix and Hulu. These libraries will license and distribute content from varied sources, but the model of broadcast television as both content producer and distributor has been shattered. In fact, the world of broadcast media delivered via cable is threatened. The threat of consumers cutting the cord, or access via cable, is real, although sports content via Disney-owned ESPN, remains within the realm of cable. World Cup Soccer fans, like myself, had to be creative in finding access to live coverage in 2014!

The four players in the entertainment world are being split and re-assembled – creators, distributors, deliverers and users.

Amidst the scramble in this colliding world, is the rise of a behemoth, Netflix, which is merging delivery, distribution and creation of content. Particularly noteworthy is the recent blurring of distinction between OTT platforms as pure distributors of content and creators of content. Amazon and Netflix distribute content over the Internet but also create content; hence they are located on the right side of Fig. 5.1. Verizon's recent purchase of Yahoo's Internet business suggests the emergence of a behemoth, where content, distribution and delivery are merged.[7] Similarly, television networks, as distributors, have ownership stakes in the content company, Hulu.[8]

Netflix is as much a global entertainment company as a technology company – it set up its own content distribution network in 2011 with servers in multiple locations. It transformed television viewing by empowering consumers with viewing options as to time and length of viewing sessions. Called streaming video on-demand (SVOD), Netflix purchased the rights to Mad Men from American Movie Classics and placed the entire first four seasons of Mad Men online – bingeing took off [65]. Early in 2016, Netflix launched its global content network in 130 new countries simultaneously [66]. With a vertically integrated supply chain consisting of global IT infrastructure (which captures the interface and delivery aspects), data to target users, and content, Netflix, with 75 million subscribers, is focused on capturing the global entertainment market – quite possibly becoming the new MGM.[9] A global premiere like Daredevil is engineered by locating content files for the movie on local servers around the world. To avoid congestion on the transoceanic cables that form the Internet's backbone between continents, Netflix has Open Connect appliances that act as a localized content delivery network at strategic, high-density locations.

Solving the content delivery problem locally leaves open the question of content itself. Regional content must have global appeal, so when "Netflix does Bollywood, for instance, it will do whatever version of Bollywood it thinks has the best chances for success not just in India, but in Arizona" [65]. Despite the powerful distribution technology, the aspiration to universally appealing content also raises something of an existential question: If Netflix engineers a Bollywood production to the point that it has universal appeal, will whatever that looks like simply be a regression to the mean? Is Bollywood via Arizona still Bollywood at all? Or does making a movie or show that's all things to all people just become mush? [65]

This relates to the issue of cultural homogeneity that I brought up in Chap. 1 and define later in this chapter.

Globalizing content faces several hurdles such as local governments, ISPs and local competition. Protectionism can keep Netflix out of a country entirely, while a lack of interest from an ISP in Open Connect can result in a jagged streaming experience. Local content creators might not want to integrate with the Netflix ecosystem as in Brazil's Globo, one of the largest commercial television networks in the world, and the largest in Latin America, which has not yet licensed any content to Netflix at all. Importantly, local video-on-demand or over-the-top streaming services are more familiar with the market and the language and have a first mover advantage.

Over-the-Top Content

Internet television or streaming television content digitally has provided an alternate channel for disseminating content and here, too, there are behemoths. The distribution of content over the Internet is termed OTT (over-the-top). The distributors are also called *edge providers*, to distinguish them from broadband providers (ISPs such as Comcast, ATT) and backbone networks (Level 3, Cogent, Akamai), which provide long haul fiber optic links and high-speed routers carrying vast amounts of data.[10] Dominance of the broadband segment by OBs such as Comcast suggest that there are significant economies of scale involving fixed costs in setting up the infrastructure – a frequent argument made by the opponents of net neutrality, an issue discussed in Chap. 7.

The unbundling of distribution from creation has allowed dissemination of content produced by independent artists, along with the traditional fare produced by the major television studios (NBC, HBO, etc.). Digital distribution channels substitute for sales in physical distribution channels such as television and downstream retailers like Walmart, who sell DVDs of television shows. Further, since pirated television content may already be available on the

Internet, streaming content may have to be underpriced to compete with free, pirated content. Importantly, streaming content may increase interest in the shows leading to legitimate purchase and decreased piracy. Increasing a product's awareness by providing content on multiple channels is like advertising, which has a cost, but the benefit is increased interest and legitimate sales [68].[11]

There are two types of OTT intermediaries: active resellers with control over the pricing and marketing details of the transactions passing through their channels, and those who simply act as a passive marketplace [68].[12] Digital downloads of content through Apple's iTunes Store or Comcast are in the category of reseller since the Apple, and not the content producer or artist, determines the terms of sale. Warner Brothers and Sony Entertainment are also resellers or distributors. Extending the reach of marketing, United Artist Entertainment, the movie studio, has ownership interests in Regal Entertainment Group, which owns the theatres, and thus monitors marketing *and* distribution.

On demand streaming services such as Spotify and Pandora for music and Amazon for streaming video are examples of the marketplace, which is a passive intermediary. The artist, as was the case with Radiohead and Taylor Swift, retains control over the streaming distribution of his or her creations and not the intermediary.[13]

For mobile service, there is the further concern of interoperability or compatibility across different OTT platforms, such as those based on the iOS operating system and the Android system. Technical compatibility across different platforms has strategic implications and therefore should be considered as more than a passive holdover from prior generations of products. It can be a strategic business choice and, in this sense, is a double edged sword. Figure 5.2 illustrates the difference between single-homing for users, where they are limited to a single OTT platform and multi-homing, where they can access both OTT #1 (iOS) and #2 (Android). Multi-homing for the users enhances their choices. But it also lowers the benefits of price decreases since the gain in demand due to lower prices is shared across platforms; this is the indirect network effect. With limited compatibility, content and apps for Android supported devices remains distinct from content for iOS supported devices. There is increased competition for users because of this all-or-nothing market. Compatibility between the two operating systems would decrease the need for market-enlarging competitive strategies since the gains from a larger user base are shared by all platforms.

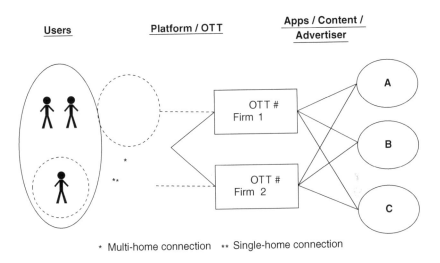

Fig. 5.2 Single and multi-homing for users

In the digital book market, the early intermediary as resellers were Amazon (launched in 1995) and Barnes and Noble (launched in 1997). Booksellers used the wholesale model where the retailer, as the intermediary, intercedes between the publisher and the consumer, who typically would pay double the wholesale price. When Amazon entered as an online intermediary, it paid the wholesale price to publishers and resold at deep discounts (often at a loss) to consumers. When Amazon introduced the Kindle in 2007, it offered bestsellers for $9.99, threatening the publishing industry. Subsequently, Apple released the iPad in November 2009 and negotiated with five publishing companies (Hachette, Harpercollins, Macmillan, Penguin, and Simon & Schuster), inducing them to join the platform by raising the prices of bestsellers to $14.99. The agency or retail model, offered by Apple in its iTunes store, is one where the publisher sets the retail price and Apple gets a 5 % commission on each sale. The Department of Justice sued Apple, along with the publishers, in April 2012 over collusion in setting the price, in violation of Section 1 of the Sherman Act. The court claimed that Apple and the publishers desired to increase the publishers' pricing power to destroy the monopoly exercised by Amazon. It is interesting to note that Amazon's below-cost pricing is itself anti-competitive since any entrant would lose money if it matched Amazon's prices.[14]

Pricing Broadband

Digital technology for the production of media content has a crucial logic of costs: the original content has large fixed costs and practically zero marginal costs of reproduction. How are the fixed costs of original production going to be covered? There are three revenue models that prevail. One is the advertiser-supported model, the second is the subscription-based model, where users pay either a fixed fee per time period or a per-usage fee and the third is some mixture of these two.[15] The mixed model is the standard one followed by most Internet Service Providers (ISP). For example, Comcast will charge a user fee plus a subscription to various individual services such as Netflix and HBO.

Internet services have two components: data usage and the speed with which this data is accessed. When access to a product can be priced separately from usage of the product, as in amusement parks, credit card services and shopping clubs, we have a pricing structure called two-part tariffs. There is an access fee plus a per unit charge based on metered usage. At Disneyland, for example, you would pay a fixed entrance fee and then pay for each attraction on usage basis. In the case of broadband, we have access to the service and a usage charge levied against data and against speed, so there can be multiple combinations addressing both components of the service.

For data consumption, Internet Service Providers sell a complex menu of options, which effectively amounts to price discrimination. In most US markets, cable and Internet are bundled commodities, marketed by the Internet Service Provider (ISP). Since the cable companies had already incurred the fixed cost of laying the cable, the marginal cost of adding Internet was small. Hence, broadband Internet access is mostly via cable and digital subscriber line or telephone technology. Fiber optic communication is a newer technology, being introduced incrementally in the US. Recent research suggests that the average customer uses 60 GB (gigabytes) of data per month. As users shift to streaming video with more capabilities, they will need more data in their home. With finite capacity, data prices will entail quantity restrictions or data caps, as is the case with wireless mobile devices. Any use in excess of the cap will be priced at higher rates. The large ISP's are currently setting data caps for streaming content obtained from sources outside the user's ISP, and "zero rating" (zero data caps) for streaming content packaged with the ISP's product, as in Comcast's Stream TV service [70].

There is concern that this pricing structure will increase the digital divide beyond simple access to the Internet. Now the cost of data could further divide the population into those who can take full advantage of high-bandwidth applications like streaming and video conferencing, and those who have to limit their usage for fear of incurring overage charges.

Another form of price discrimination, known as commodity bundling, is the practice of selling multiple distinct channels as a cable package, for a single price. Users might be discouraged from cord cutting, or cancelling their cable TV package, by pricing the a la carte components in a specific configuration so as to induce them to purchase the prix fixe menu, even if many components of the bundle are not consumed. The cable package benefits from direct and indirect network effects when cable and Internet service come from the same provider. For example, if Netflix lowers its prices, more users are attracted to Comcast's service, benefiting Hulu as well. There is also the direct network effect as when Netflix increases its content library, more users are attracted to the platform, which in turn incentivizes Netflix to add yet more content [71].

How is speed component of Internet service priced? Data usage caps apply to the size of the service consumed, but the speed is sold in discrete increments. For example, in the Princeton, N.J. area, Xfinity cable by Comcast offers a speed of 25 Mbps for $39.99 and 150 Mbps for $82.95 with no data cap. Verizon Fios offers fiber at 100 Mbps for $59.99, 150 Mbps speed for $69.99 but a highest speed of 500 Mbps for $269.99/month.

What about high-speed broadband connections? The technology is available, but implementation costs are the bottleneck. To replace existing broadband connection, that is cable or DSL, with fiber involves enormous fixed costs. The OTTs are now entering the delivery business with Google Fiber already available in Kansas City, Austin, Provo, and Atlanta. Google is developing an innovative strategy by adding high-speed fiber to existing infrastructure. This obviates the more intensive infrastructural investment while simultaneously delivering the speed. For example, Google has partnered with the company Webpass, which already beams the Internet to a fixed antenna on the building. Google then runs data cables into each unit, offering speeds of 1 Gbps [72].

Net neutrality requires all content to be treated equally by the ISPs but access to this content is still subject to differential pricing, and as mentioned above, increases the digital divide. Broadband pricing comprises a price for bytes of data transferred and an associated speed of data transfer. Even if the price per byte is constant, the speed can be throttled or slowed down, effectively compromising the service. If, for example, Comcast throttles the data transfer in some parts of its network, then there is unequal access to content. Since information and entertainment are frequently joint products, unequal access to information can this move the needle on knowledge-generated social and economic inequality.

SOCIAL MEDIA AND SOCIAL NETWORKS

Social Media

Social media (SM) are one of the spinoffs resulting from the unbundling of the medium and the message in the media industry. SM are the platforms via which communication takes place and the content is derived from the consumers in social networks (SNs) through person-to-person communication, for the exchange of information and general socialization. In distinguishing the two, I want to draw attention to the difference between the organization or medium (SM) that manages the network of individuals and the network itself which produces the content.

The revenue model for SM is advertising based. Consumers spent 25 % of their time on mobile versus 22 % on desktop in 2015, but advertisers allocated only 12 % of their ad budget to mobile compared with 23 % to desktop advertising, so there is enormous growth potential [73]. In a prescient business move, in the third quarter of 2015, when the number of monthly active users on Facebook was 1.65 billion and 54 % of them were mobile users, Facebooks' mobile advertising accounted for 78 % of advertising revenue during this same quarter [74].[16]

SN engagement as measured by reach and digital audience penetration shows that Facebook dominates both average monthly minutes per visitor (over 1000 minutes) and percent reach among eighteen to thirty-four year olds (close to 100 %), with Snapchat and Instagram equal in engagement but Instagram ahead with more than 60 % reach. The leaders in messaging are WhatsApp, Facebook Messenger and WeChat [75].

SM have enabled the development of granularity in the CIME industries by addressing content issues such as (1) information, and voice to any individual, thereby enabling granularity in information provision, (2) entertainment, and (3) collective action facilitators, as in the political arena. SN satisfy individual social imperatives such as (4) engagement, (5) empowerment and (6) immediacy. The consequence is a (vii) global citizen. I discuss each of these in turn.

(1) Information

The SM platform is a granular form of connectivity since it is the digital equivalent of word-of-mouth communication. The critical differences are the massive and simultaneous reach of digital word-of-mouth. Text messaging and postings on Facebook pages, where individual voices have a public canvas, are examples of granularity, and there is no more forceful example than the political arena. Forrester Research calls this the age of the consumer, and predicts that "empowered customers are changing the fundamentals of the market" [76].

Word of mouth has never been so important or so instant. From Instagram and Pinterest to social sites such as Snapchat, a consistent theme is instant reaction to people, products and ideas. Snapchat attracts 41 % of millennials in the US and is aiming to be the leader in live event socializing on mobile devices, as was the case during the 2016 Rio Olympics when users viewed the events on *Live Stories* [77]. Consumers are connecting to strangers, becoming active content creators and sharers. Communication about a product, say a new movie or a sports event, can now be instantaneous and disseminated to large numbers of consumers, leading to viral connectivity. Advertisers now use the power of WOM communication, since consumers filter out general media advertising.

When users stream live broadcasts, using their smartphone video apps, of charged political events on Facebook Live, as part of their News Feed page, they are substituting for the news media at large. The content is unfiltered and possibly incendiary. The perspective on these events is localized, providing superior insights compared to live broadcasts, but might also be biased. Started in the spring of 2016, Facebook Live is in a partnership with *New York Times* for creating live broadcasts, so it is possible that the Times editors would provide the curation.

Information put forth on SM can acquire commercial aspects. Facebook "likes" and product reviews are a form of unpaid advertising. There also exists a paid version of this form of publicity, where a reviewer gets paid by the business to publicize their appreciation of its product. In this instance, the distinction between the OTT #1 and advertiser A in Fig. 5.2 becomes blurred as Facebook integrates along the supply chain.

(2) Entertainment

SM functions as a platform for delivering the message, which is entertainment in addition to communication and information. Information and communication may have been the original intention of this platform, but

SM perusal during the prime time television viewing hours suggests that entertainment is an important element. For example, time spent on SM may decrease the time spent on watching television programming, although Liebowitz and Zentner find that Internet adoption has had a minimal impact on television consumption [78].

The National Basketball Association (NBA) has more than one billion followers on SM accounts in the US and on Tencent and Sina Weibo in China, covering team, player and league accounts, more than the National Football League, Major League Baseball and the National Hockey League. SM provide the perfect virtual WOM that is a quintessential part of the sports audience. There is a tremendous spillover effect from SM to regular broadcast television as the SM chatter leads individuals to view the live televised event. A Nielson study, commissioned by Facebook, of social conversation about nine NFL games, "found that each additional share of a Facebook post about a game in the 15 minutes before it started correlated with an extra 1,000 viewers for the first minute of the broadcast. That's not a threat to traditional broadcast; it's a lifeline" [79].

(3) Collective Action

BlackLivesMatter is an SN, and a social movement in that it embodies a philosophy. It became a network under that hashtag in 2013 when Alicia Garza, a civil rights activist in Oakland, California published her opinion of the acquittal of George Zimmerman, who was responsible for the death of Trayvon Martin. Social movements need technology to generate collective action. In 1965 Alabama, while the telephone lines were heavily used, the most important technology was the

> set of film canisters being ferried past police blockades on Highway 80 by an ABC News TV crew, racing for the Montgomery airport and heading to New York for an evening broadcast. That night, 48 million Americans would watch the scene in their living rooms [80].

Martin Luther King Jr. would go on to use television as his technological tool.

The BlackLivesMatter movement is a decentralized structure, helped by a plethora of platforms, depending upon the objective. Information is the institution supporting the movement and posting videos of police violence on Instagram, Periscope, Vine or Facebook, acquires instant virality. However, the diversity of platforms dilutes the strength of the message. Even with the heightened visibility afforded by SM, the deeper message of the movement needs to be constantly reaffirmed [80]. SM gives a voice to any individual, regardless of demographic category, and propagation of this voice is by friends so no organized publicity machine is required.

At a pragmatic level, citizens of the small town of Jun, Spain (pop 3500), where more than 50 % of the residents have Twitter accounts, demonstrate participatory democracy: they communicate with local government officials about the provision, or breakdown, of public services via their Twitter accounts. Twitter communication is highly visible so tweets made about a broken street lamp are immediately followed by the mayor's response which includes the electrician's name. The free publicity to the electrician is paired with a reputation-on-the-line threat, so the lamp is fixed in less than twenty-four hours [81].

In the 2010 election, Facebook reminded voters to go to the polls, resulting in 340,000 additional votes nationwide. The company's get-out-the-vote strategy in 2012 was not as successful due to software bugs. In early 2014, Facebook altered the emotional tone of 500,000 users' News Feeds to see how that changed their subsequent posts. In the 2014 mid-term elections, all Facebook users logging onto the site on Election Day were reminded to vote and an "I'm a Voter" button was displayed on the Facebook page. Facebook has displayed the "I'm a Voter" button in ten national elections, including those in India, Brazil and Indonesia [82].

On the negative side, governments are concerned about the use of SM by terrorist organizations. Robert Hannigan, the head of Britain's intelligence agency Government Communications Headquarters (GCHQ), has "castigated the giant American companies that dominate the Internet" for providing the vehicle for terrorists to advance their cause. Mr. Hannigan calls for "a new deal between democratic governments and the technology companies in the area of protecting our citizens," in a balance between national security interests and civil liberties [83].

Social Networks

The SN of connected individuals facilitated by the SM platform is characterized by engagement, empowerment and immediacy, but results in a global citizen.

(4) Engagement

The platform provided by SNs is a passive intermediary, but it is one that supersedes all other historical intermediaries that played the role of bringing communities together. Whereas groups of people would congregate at the water cooler or the mall, these groups tended to be small and less connected. There were multiple small groups congregating at water coolers and malls-as-town-squares across the world. One would not consider the mall as an intermediary, but it performed the function of bringing the community together. Modern SNs scale up the idea of the mall-as-town-square by bridging global communities, eliminating the smaller intermediary town-squares. A reunion of college classmates can be facilitated by Facebook without the intervention of a Reunion committee at the college level.

The discussion in this book pertains to connections at the tangible level. One could imagine connectivity as a layered cake, so that we are addressing only the icing on the top and the essential elements are in the cake below. As Sherry Turkle suggests, the connectedness we feel as a result of this technology is in a sense illusory, it is an "empathy machine." The empathy that is present in face-to-face conversation is absent from virtual connection. Effective communication consists of both connection and conversation [22]. Perhaps there are deeper layers of connectivity that address emotional aspects, beyond the economic and social ones.

The content on SM is provided by the users. However, there is additional content curated by algorithms developed by the platform, using personal preferences revealed by users. For example, the News Feed on Facebook, provides developing stories of interest to users with a given profile. Clearly, as users update their profiles with likes, their News Feed will also get updated. The trending topics box is another source of general news, although relatively minor on mobile devices. Therefore, the notion that the platform is a gatekeeper of news is misleading. By applying user information to the sorting and filtering of news, the platform acts as an ostensibly neutral curator. However, to the extent that the criteria for sorting are themselves decided by the platform, news can be tailored to various demographic groups on SM, violating the notion of strict objectivity.

More insidious is the feedback loop since user reaction to News Feed and trending topics impacts the algorithms, which are then tweaked so the revised News Feed falls in line with our new user profile. Algorithms are complex computer programs, created by programmers writing millions of lines of code, using data from economic and social connections. Machine learning, where computers create patterns from masses of data, allows the content to be updated. As we consume news and other content, our reaction to the News Feed and trending topics is mined by algorithms for patterns, which influence the content, which update our reactions, and so on. The public policy question here is how these patterns update content, an issue I return to in Chap. 8.

(5) Empowerment

An SN is a set of linked nodes empowering the user with a shared identity with communication and other organizational benefits across the network. Each individual's page leaves a digital footprint by providing information about name, profile, availability, location and real-time updates about the person. Psychologically meaningful links can be drawn between the user's personality and the information provided on the user's SN page. Two features of SN are critical.

One, there is an implicit contract of responsibility and trust between members of a SN; responsibility for maintaining reputation and credibility, while enjoying the benefits of trust. A possible threat to this implicit contract is virulence created by a single comment due to the velocity and direction of the conversation. A single unpleasant rumor about some user could be casually disseminated and then acquire a force of its own through information cascades.

Second, while a SN provides a platform for sharing ideas, photos and products, how much of your identity is "owned" by the SN? Even after you delete your account, the SN has already monetized your information. Texting and tweeting are very public and transparent forms of connections. There are benefits from this transparency, particularly in the public sphere, as we shall see in Chap. 5.

What is the logic behind sharing? Why do people want to share pictures, videos, articles, pages from books? By sharing pieces of one's identity, one is connecting in a controlled manner – the persona shared is carefully curated and published at the discretion of the owner. Sharing is a form of acknowledgement, as one human being acknowledges the status or worth of another human being. We also share to satisfy the need for, and affirmation of, kinship and reciprocal altruism. Both are biologically grounded, according to Fukuyama, as it is not merely survival of the fittest organism but rather "survival of that organism's genes." By exchanging personal moments of one's life there is confirmation of membership in a common tribe. Individualism, and the drive for privacy, as we shall see in Chap. 7, is the core of our political and economic models today, only because, as Fukuyama writes, "we have institutions that override our more naturally communal instincts."[17]

On the more somber side, this sharing could lead down malevolent or undesirable paths when privacy and self-promotion become the dominant features. Here again, adequate systems of trust and responsibility are vital. Much of the economics of SM rests on good governance and pro-social preferences. Dixit explains, "A society whose members have such preferences can take collective action that benefits the whole group" [33]. These are societal norms developed from internal value systems that preclude online malevolent behavior.

(6) Immediacy

Messaging is a manifestation of connectivity is daily life and in real time. Messages can be sent in two technical formats: SMS or short message service transmits data in the form of texts over a regular cellular phone service from one phone to another, and doesn't require synchronicity. IM or Instant Messaging sends messages over the Internet, and, in the days before smartphones, required both parties to be online simultaneously.[18] With the advent of smartphones, IM advanced to consecutive entries rather than simultaneous chat.

Nearly 75 % of the adult US population will use smartphones compared with less than 50 % print newspaper readers in 2016. According to survey data from the marketing firm eMarketer, of the time spent during the day on various smartphone activities, 22 % was devoted to entertainment and 65 % was allotted to communication (of which an equal amount, 22 %, was spent on texting and phone, with 10 % each on email and SM) [84]. How do users decide upon the appropriate medium of communication when choosing between texting and phone? The spontaneity and simultaneity offered by phone is deceptive since the call may simply be switched to voicemail, which entails a lag that can exceed that of messaging.

Recent messaging apps, such as Signal Private Messenger, Gliph, Wickr, Telegram and iMessage have sophisticated encryption capabilities, with some apps (such as Telegram) having the capability of programing messages to self-destruct from both devices at a certain time. Snapchat, the picture-messaging app, features user-generated videos, which creates a community of viewers since all are watching the same videos. In the world of entertainment, messages on Twitter alert viewers to movie and shows, and the "fear-of-missing-out" or FOMO persuades even more users to gravitate toward that form of entertainment.

Messaging is becoming the primary form of communication with more than 1.6 billion people worldwide, or 22 % or the global population, being regular users of mobile messaging apps. This has allowed firms, who have a presence on the app, to offer customer service for individuals on their messaging platforms using "bots," which are algorithms using artificial intelligence for real-time conversations about products and services. Communication and shopping are merging onto a single platform as in Tencent QQ, the instant messaging platform and WeChat, the mobile chat service, both owned by Tencent, the Chinese Internet holding company.

Beyond the world of personal memorabilia, SNs are critical in the spread of ideas. How do ideas spread through networks? Discussion on the SNs provides insight into the zeitgeist or culture of the age. The structure of the network (pattern of links) is important not only for opportunities to transact but also for the diffusion of ideas and innovations. For a new idea to catch on, one can argue that a critical minimum fraction of individuals in the network must switch to this idea. If the idea remains isolated with a single individual or an individual who is only connected to, for example, two other individuals, then it has not reached a critical threshold so the idea disintegrates.

(7) Cultural Homogeneity or the Global Citizen

More connectivity, empowerment and immediacy support the assertion by historian Francis Fukuyama that connectivity is taking us into human socialization beyond kinship ties, religion and national identity. Culture is the set of norms that determines which ties are formed, whether they are strong or weak and directional. The global democratization of culture, or cultural homogeneity, is made possible by digital connectivity. The search for individual identity then becomes the major force driving political change. The recent UK referendum on its membership of the European Union is an example, with 52 % of the population voting to exit the EU.

SNs have a homogenizing influence that we now recognizing. Western global diaspora is captured by the cosmopolitan citizen when meritocracy captures the top ranking students from everywhere and "homogenizes them into the peculiar species that we call 'global citizens'." This form of "elite tribalism is actively encouraged by the technologies of globalization, the ease of travel and communication.[19] There is more genuine cosmopolitanism in Rudyard Kipling and T. E. Lawrence and Richard Francis Burton than in a hundred Davos sessions" [85]. Cultural homogeneity is best defined by the description given to these elites by Ross Douthat:

> They have their own distinctive worldview (basically liberal Christianity without Christ), their own common educational experience, their own shared values and assumptions (social psychologists call these WEIRD – for Western, Educated, Industrialized, Rich and Democratic)...And like any tribal cohort they seek comfort and familiarity: From London to Paris to New York, each Western "global city" (like each "global university") is increasingly interchangeable, so that wherever the citizen of the world travels he already feels at home [85].

ADVERTISING

Advertising is a major player in the entertainment world, sustaining two of the three revenue models. "In the national accounts, a business whose revenue comes entirely from advertisers is treated as providing all of its services to ad buyers" [49]. This classification would imply, in the context of our three-party model in Fig. 5.2, that consumer welfare is not counted in the value added to gross national product by content producers. The media, much like broadcast television has been doing, sells advertisement services to pay for content, making their services an intermediate input. Consumer attention is being sold to the purchasers of advertisement space, with the transaction patched together by media firms, who themselves pay nothing or buy this attention at zero cost. In fact, *consumers* pay a low fee (and often it is free) for media content even as they are freely giving their private information to the very same firms. Under the same logic, "many internet services have that same treatment: Facebook and Google are counted as providing advertising services to businesses not services consumed by households" by national income accounts [49].

Economic theory posits consumer demand as being derived from fundamental preferences. However, what do these preferences actually cover? Is it functionality, that is, "what does the product do," or is it the product itself? What should advertising properly target – product categories or product functionality? The former is conventional advertising that follows standard demographics such as age, income and gender. The latter targets people who would naturally respond to what the product does.

There are two types of advertising: informative advertising and persuasive advertising.[20] Informative advertising increases demand by providing basic information about the product. This type of advertising works well for search goods, where quality is determined prior to purchase. Advertising could also provide complementary information such as prestige value by revealing consumption preferences of celebrities and major public personalities. Demand could be impacted in two ways. One, demand can increase (the demand curve can shift outward in the standard price-quantity graph) due to greater information about the product or the increased sales of complementary products. If iPhone sales increase, it is quite possible that iTunes sales will also rise since the two are paired products. Two, demand can be created if new product categories are created, such as the iPad. Informative advertising is the standard advertising model where businesses draw price sensitive buyers by pricing low and getting a large number of initial buyers.

Persuasive advertising is more relevant to the world of SM since businesses can leverage the gatekeeper function of fashion leaders or influential people by persuading them to buy a product. These influential people are not necessarily those who have lots of friends or are part of a network with a high percentage of connected neighbors or networks with a high clustering coefficient, but rather individuals who act as gatekeepers between highly embedded SNs. They can be the seed of a chain reaction in a network if individuals only care about their immediate neighbors' purchases in a sequential decision process. From a strategy perspective, it is important to note that SNs exhibit some patterns. People tend to have common friends so there is some clustering in the network space. Advertising to this cluster is wasteful since the connections within the network will promote the product via WOM. Advertising to multiple gatekeepers within the larger network leads to a greater probability of covering the entire network since they act as seeds for information cascades and network effects.

Mobile Advertising

Ads-in-the-moment are the fastest growing segment of the advertising industry. This is the mobile advertising industry, which relies on spontaneity and instant reaction. Persuasive advertising bends the demand curve by creating genuine (or spurious) product differentiation and brand loyalty. Mobile advertising promotes impulse buying, which is captive to persuasion so that mobile devices can be strong channels for persuasive advertising. When consumer preferences are malleable, they can be made less sensitive to price increases, thereby increasing the profit margin. Branding and loyalty programs seek to follow the latter strategy.[21] Hence, the goal of advertising is not only to increase demand but also to increase brand loyalty. Experience goods, where quality is determined only after purchase, are more susceptible to persuasion. Consumers could be persuaded to believe that there is a better match between product and buyer relative to other goods. Or they could be reminded of a previous experience in order to generate repeat business.

There has been an increasing interest in "event television programs," like sports, which draw advertisers because viewers watch the program live, as in the 2014 Soccer World Cup. Viewers must see the ads – there is no option to skip the ads as in more general media programming. Increased participation in SM has generated significant increases in marketing budgets devoted to digital media in general, and mobile devices, in particular. The combination of event advertising and mobile devices is powerful since attendees at the live event will likely be connected to their devices, making them perfect targets for mobile advertisements while watching the live event in person.[22]

Reviews and Ratings

Reviews on SNs work by individuals voluntarily sharing information or, more recently, by firms paying users to "like" their product. The network then becomes a high-speed WOM medium, which is very potent in the advertising world. Social advertising is a term that applies to both forms of customer reviews, voluntary and paid, since they rely on TRR&R – this is the glue that holds the network together. An online ad that the consumer has agreed to display on an SN, along with his/her picture and name, can be a significant development for advertisers since they can tap into an individual's SN for targeted advertising. The introduction of improved privacy controls can result in users being twice as likely to click on personalized ads. User-generated content makes the ad more persuasive (people are more likely to buy) and also have high reach (appeal to more consumers), since people are engaged by these ads and thus are sharing them as well. However, ads must not be obviously commercial when exploiting SM. While sharing of ads may increase, the likely provocative nature of the ads will reduce its persuasive content. Tucker feels that humor and visual appeal are more persuasive than outrageousness, sadness or anger [87].

Customer reviews represent free advertising, but they are also optional viewing. They generate collective wisdom due to pooling or crowd sourcing of all past users of the product or service but what is the incentive for users to write reviews? Social influence is thought to be the most important factor inducing reviewers, but they could be misleading as they may simply follow previous reviews, thereby generating an information cascade rather than being independent sources of information or representing true beliefs. Reviews are also time sensitive, since the product may change and earlier reviews become irrelevant. The salience of the review or information is also important. Salience represents the ease of inferring relevant information, so that a lengthy review, which addresses marginal concerns of the user, will be disregarded. Numerical reviews raise another concern. These are reviews where users assign "stars" or a numerical score to a product. Does a product with five stars from ten reviewers outrank one with two stars from fifty reviewers? Would it matter if the latter product had reviews dated six months ago while the former's reviews were more recent?

Some firms can game the system by paying certain consumers to write favorable reviews. Mayzlin et al. [88] find that fake or promotional reviews are more likely when the firms are independently owned and smaller compared to large chains.[23] Using data on TripAdvisor and Expedia they find that an independent hotel owned by a small owner will generate an incremental seven more fake positive TripAdvisor reviews than a chain hotel with a large owner. This suggests that a chain hotel that is located next to an independent hotel will have six more fake negative TripAdvisor reviews compared to an isolated chain hotel. The reputational benefit of posting a fake review accrues to only one hotel (the one posting the review) while the cost of posting the fake review (getting caught) multiplies in the number of hotels. This explains why smaller entities have greater incentives to post fake reviews. Further, smaller firms are more likely to be owner-operated so their residual claims on profits are larger than for employees of larger firms, who have no residual claims, but instead have a fixed salary.

User reviews matter less when preferences are heterogeneous. Using data from the iPhone ecosystem, Yin et al. find that previous app experience but no updating incorporating user reviews, enables a game app to become a killer app, while the opposite is true of non-game apps [89]. Ease of entry into the game app industry makes it vital for the new entrant to carve out demand the first time.

On balance, how has digital and communication technology (DCT) altered the world of advertising and media content? Replication and personalization of ads made possible by digitization has reduced the tradeoff between customization and mass marketing so product information is targeted. Before the commercialization of the Internet, firms had to choose between personal selling, which is an incredibly rich form of marketing communication but has limited reach since there are no economies of scale, and media or television advertising, which achieves impressive reach but is not a rich form of marketing communication. Now, firms can harness real-time marketing by following conversations, participating as in a social gathering, staying alert for new developments such as popularity of a new Twitter account so they can quickly tailor ad content to real-time news. Online reputation management is a powerful strategy where firms encourage new customers to write reviews and actively monitor reviews.[24]

My Take

While there are strong forces toward OB there are stronger currents limiting branding. Persuasive advertising is unlikely to sustain the powerful brand. Customer reviews are transmitted faster than objective information and this reduces the power of brands as consumers are more willing to rely on friends' reviews than advertising. Despite information cascades spreading reputation nearly instantaneously via viral marketing, network effects adding positive feedback, and older brands having first mover advantage, there is a meaningful change in the mode of consumption.

Two significant developments have occurred: renting and fragmented consumption. Both result in demand for temporary and unbundled access to content. These developments preclude sticky consumption as buyers will sample newer products in a quest for novelty. The desire for autonomy, for the freedom to make un-coerced decisions, is surfacing, as we saw in labor markets and in the market for individuals' attention, and buyers may resist being pigeon-holed into well-defined categories (brands).

Notes

1. Performance rights are from broadcast and personalized streaming services, while synchronization revenues derive from the use of content in film, television and brand partnerships.

2. Apple's streaming music service will showcase new albums, via iTunes, before they are available on other providers, but unlike Spotify, it will have no free tier. Some artists own their streaming music platform (rapper Jay-Z owns the music platform Tidal), while others temporarily keep new releases off streaming services to maximize more lucrative CD and download sales. In 2014, artist Taylor Swift made her albums available on Tidal but not on Spotify [61].

3. BitTorrent is the peer-to-peer file sharing software, developed in 2001, that enables users to transfer digital music files to one another at practically zero cost. It is based a file format founded by MP3.com in 1997, which was bought by Apple in 2000 as a basis for iPod, the portable mp3 player, and later, iTunes.

4. Adele's song "Hello" from her new album *25* obtained a record breaking 1.11 million download sales in a week, upon its release on October 23, 2015. [60]

5. The infrastructure I refer to is the backbone network and the content delivery network, comprising four modules: physical layer (copper wires, fiber); the link layer; the network layer and finally the transport layer. The final module, over the top of this stack of four is the application layer, which is our focus. See the discussion in Chap. 3 on Internet Architecture for more detail.

6. Modeling the content consumption stream as a perpetuity, the present discounted value of this stream is given by:

$$PV = \frac{\text{subscription fee}}{r} \text{ where } r \text{ is the discount rate.}$$

7. Yahoo sold its search, news, sports, email and Tumble SN to Verizon for $4.8 billion as of July 25, 2016.

8. Hulu.com is an advertising supported Internet portal for streaming video content produced by network television. Television networks themselves took the initiative in creating this platform and it was launched in March 2008 as a joint venture between Fox and NBC Universal. In April 2009, Disney/ABC reached an agreement to take a partial ownership position in Hulu.com and add its content to the site. In a move toward integration, Comcast, the ISP, has also partial ownership in Hulu.

9. MGM or Metro-Goldwyn-Mayer is a legendary movie studio, founded in 1924, and a behemoth in its time.

10. These are the definitions used in the recent statement made by the D.C. Court of Appeals in the net neutrality judgment [67].

11. The convergence of the new and old television worlds is illustrated by the shift in the rivalry between HBO and Netflix. In 2011, some TV executives were skeptical when Reed Hastings, Netflix's chief executive, pointed to HBO as his company's main rival, and HBO rejected the comparison [68].

12. More precisely, Hagiu and Wright explain that the critical distinction between the two is the "allocation of non-contractible decisions" such as pricing, marketing, customer service [68]. The reseller retains control over the marketing and distribution process, while the passive marketplace entails zero control.

13. Apple's App store, Best Buy electronics store, most department stores are examples of the marketplace form of intermediary.

14. According to dissenting Circuit Judge Dennis Jacobs, "Although the major publishers believed Amazon's below-cost pricing was "predatory," each publisher understood that it was powerless to take on Amazon. Publishers feared that Amazon might "compete with publishers by negotiating directly with authors and literary agents for rights" and might "retaliate" against insubordinate publishers "by removing the 'buy buttons' on the Amazon site that allow customers to purchase books…or by eliminating [a publisher's] products from its site altogether." Assenting Circuit Judge Lohier argued that "[i]t cannot have been lawful for Apple to respond to a competitor's dominant market power by helping rival corporations (the publishers) fix prices, as the District Court found happened here. However sympathetic Apple's plight and the publishers' predicament may have been, I am persuaded that permitting "marketplace vigilantism" would do far more harm to competition than good, would be disastrous as a policy matter, and is in any event not sanctioned by the Sherman Act." [69]

15. The industry acronyms are SOVD for subscription-based online video distributor services such as Netflix; AOVD for free, advertisement-supported services such as Crackle and Hulu; and TOVD for transactional services such as Amazon Instant Video.

16. According to the International Telecommunications Union, there are 3.2 billion Internet users globally. If Internet users in China and Russia are excluded, has Facebook, with 1.65 billion monthly active users, reached some glass ceiling in number of Facebook users?

17. He goes on to argue that "there was never a period in human evolution when human beings existed as isolated individuals" [Chap. 3].

18. In the absence of Internet via Wi-Fi or 4G cellular connections offered by the mobile phone company messages will be sent via SMS on the phone carriers network.

19. The Chinese messaging site, WeChat, covers the global Chinese diaspora and comprehensively covers the functions of all popular apps in the US such as Paypal, Facebook, Amazon, Uber, Yelp, Expedia, Spotify, Tinder and others.

20. Search engine marketing has several circularity issues. Lewis et al. [86] argue that there is an identification problem here due to the endogeneity of advertising. "First, many models assume that if you do not click on the ad, then the ad has no effect on your behavior... Online ads can drive offline sales, which are typically not measured in conversion or click rates... Second, many models assume that if you do click on an ad and subsequently purchase, that conversion must have been due to that ad. This assumption seems particularly suspect in cases, such as search advertising, where the advertising is deliberately targeted at those consumers most likely to purchase the advertised product and temporarily targeted to arrive when a consumer is performing a task related to the advertised good."

21. The profit maximizing condition calls for equating marginal revenue to marginal cost or $MR = MC$. MR can be shown to equal $P + QdP/dQ$ or $P(1 + 1/ed)$. Setting this expression equal to MC yields an expression relating profit margin to price elasticity of demand: $(P - MC)/P = -1/ed$. The lower the price elasticity, ed, the higher the profit margin.

22. The Apple Macintosh commercial during the 1984 Superbowl is famous and considered one of the best commercials. Mac was incompatible with all other computers so software written for Mac could only work on Mac and therefore needed a critical minimum network of buyers for software to be written. The 1984 commercial did not simply inform each viewer about the Mac – it told each viewer that many other viewers were also being informed about the Mac. This meant that the potential market was huge since the network effects could be dramatic and instantaneous. All who saw the ad were potential customers and part of the network, which was the entire television audience.

23. There are three characteristics that distinguish hotels: (1) size or single versus multi-unit; (2) affiliation – part of a chain or independent and (3) ownership – owner or hotel management company.

24. More recently, the Internet Service Providers (ISPs) are following this strategy of tagging users when they log on. For example, Verizon inserts a distinct and anonymous identifier into a user's online activity and then groups these identifiers into demographic categories. When a particular ad is to be shown on a website, these categories provide targeted advertising possibilities. Marketing and government agencies "could stitch together someone's anonymous identifier with web cookies to create a detailed profile that follows the person's web-browsing activities, even after Verizon generates a new anonymous identifier for the user" [91].

BIBLIOGRAPHY

[59] *Global Music Report: Music Consumption Exploding Worldwide,* Accessed June 3, 2016 from http://www.ifpi.org/downloads/GMR2016. pdf and http://www.ifpi.org/facts-and-stats.php

[60] *Global Music Report* 2016, International Federation of the Phonographic Industry. Accessed June 6, 2016 from http://www.billboard.com/articles/columns/chart-beat/6777905/adele-25-sales-first-week-us

[61] Steel, Emily, "Netflix, Amazon and Hulu No Longer Find Themselves Upstarts in Online Streaming" *New York Times,* March 24, 2015

[62] Waldfogel, Joel, "*Digitization and the Quality of New Media Products: The Case of Music,*" Working paper, August 2013

[63] Bruni, Frank, *New York Times,* June 5, 2014

[64] Kotrous, Michael, and James B. Bailey, Gigabit Fiber Networks & Speed Competition in the United States: Cross-Sectional Analysis at the Census Block Level (May 18, 2015). Available at SSRN http://ssrn.com/abstract= 2607729 or http://dx.doi.org/10.2139/ssrn.2607729

[65] Nocera, Joe, "Can Netflix Survive in The New World It Created," *New York Times Magazine,* June 19, 2016, Accessed June 19, 2016 from http://www.nytimes.com/2016/06/19/magazine/can-netflix-survive-in-the-new-world-it-created.html?hpw&rref=magazine&action=click&pg type=Homepage&module=well-region&region=bottom-well&WT.nav= bottom-well

[66] Accessed March 29, 2016 from http://www.wired.com/2016/03/net flixs-grand-daring-maybe-crazy-plan-conquer-world/

[67] Accessed on June 15, 2016 from https://www.cadc.uscourts.gov/inter net/opinions.nsf/3F95E49183E6F8AF85257FD200505A3A/$file/15-1063-1619173.pdf

[68] Hagiu, Andrei, and Julian Wright, "Marketplace or Reseller," *Harvard Business School Working Paper,* 13-092, January 31, 2014.

[69] *United States of America v. Apple, Inc., Docket #: 13-3741cv,* decided June 30, 2015.

[70] Finley, Klint, "Sorry, It's Time to Start Counting Gigabytes at Home, Too," *Wired Business,* June 1, 2016. Accessed June 24, 2016 from http://www.wired.com/2016/06/sorry-time-start-counting-gigabytes-home/

[71] Whinston, Michael, "Tying, Foreclosure and Exclusion," *American Economic Review,* 1990, no. 80. http://www.ifpi.org/facts-and-stats.php. Accessed 10/30/2014

[72] Finley, Klint, "Google Fiber Just Swallowed Up Another Internet Provider," *Wired,* June 23, 2016. Accessed on June 24, 2016 from http://www.wired.com/2016/06/google-fiber-just-swallowed-another-internet-provider/

[73] Johnson, Lauren, "7 Big Trends That Are Shaping the Future of Digital Advertising," *Adweek*, June 1, 2016. Accessed June 10, 2016 from http://www.adweek.com/news/technology/7-big-trends-are-shaping-future-digital-advertising-171773

[74] Faecbook website, Accessed July 15, 2016 from http://newsroom.fb.com/company-info/

[75] Peterson, Tim, "Facebook's Ad Volume Has Grown for the First Time in Two Years," January 27, 2016. Accessed July 3, 2016 from http://adage.com/article/digital/facebook-q4-2016-earnings/302378/

[76] https://solutions.forrester.com/aoc-predictions. Accessed November 5, 2015.

[77] Kuchler, Hannah, "Snapchat Strikes Olympic Gold," *Financial Times*, August 15, 2016. Accessed August 15, 2016 from https://www.ft.com/content/c463c9fe-6279-11e6-8310-ecf0bddad227

[78] Liebowitz, Stan, and Alejandro Zentner. Clash of the titans: Does Internet Use Reduce Television Viewing?. *The Review of Economics and Statistics* 94, no. 1, 2012.

[79] Mcclusky, Mark, "Techies are Trying to Turn the NBA Into The World's Biggest Sports League," *Wired*, May 31, 2016. Accessed June 1, 2016 from http://www.wired.com/2016/05/how-tech-took-over-the-nba/?mbid=nl_53116

[80] Stephen, Bijan, "Social Media Helps Black Lives Matter Fight the Power," *Wired*. Accessed July 16, 2016 from http://www.wired.com/2015/10/how-black-lives-matter-uses-social-media-to-fight-the-power/

[81] Powers, William, and Deb Roy, "The Incredible Jun: A Town that Runs on Social Media."Accessed June 7, 2016 from https://medium.com/@social machines/the-incredible-jun-a-town-that-runs-on-social-media-49d3d0d4590#.ar68xkmt8

[82] Goel, Vindu, "Facebook Has Its Own Get Out The Vote Message," *New York Times*, November 5, 2014

[83] Cowell, Alan, and Mark Scott, "Top British Spy Warns of Terrorists' Use of Social Media," *New York Times*, November 5, 2014

[84] eMarketer, Accessed July 19, 2016 from https://www.emarketer.com/corporate/coverage/be-prepared-mobile?ecid=m1216&CTA&mkt_tok=eyJpIjoiWTJFek16UmlORGMwTVdFMSIsInQiOiJ1YXE2SmFcL2pWeVI4d1NyRHJnQ1BvVmtQVGR0WlJaRUJTd21zVzNEZlNySnFZNFpMeU9COGNyWXMwNUJBNFwvRWVVQRmFzdk5oUVFmZnI5Y1QwQURhZ2t5YjZlSExxZWYzN1Azc0hlwV2M0Sjc0PSJ9

[85] Douthat, Ross, "The Myth of Cosmopolitanism," July 2, 2016, Accessed on July 15, 2016 from http://www.nytimes.com/2016/07/03/opinion/sunday/the-myth-of-cosmopolitanism.html

[86] Lewis, Randall, Justin Rao, and David Reiley, "Measuring the Effects of Advertising: The Digital Frontier," *Microsoft working paper*, May 2014.

[87] Tucker, Catherine, "Ad Virality and Ad Persuasiveness," January 2014, http://ssrn.com/abstract=1952746

[88] Mayzlin, Dina, Yaniv Dover, and Judith Chevalier. Promotional Reviews: An Empirical Investigation of Online Review Manipulation. *American Economic Review* 104, no. 8, 2014.

[89] Yin, Pai-Ling, Jason Davis, and Yulia Muzyrya. Entrepreneurial Innovation: Killer Apps in the iPhone Ecosystem. *American Economic Review* 104, no. 5, 2014.

CHAPTER 6

The Sharing Economy: Information Cascades, Network Effects and Power Laws

Sharing can morph into copying as nodes imitate each other's benign behavior. This can have adverse system wide effects.

Abstract Connectivity generates familiarity and induces collaboration, and also the phenomenon of copying, known also as FOMO (fear of missing out). Imitating others' behavior derives from the social influence of networks which is different from homophily or joining groups with similar characteristics. While financial markets have manifested granularity in the unbundling of functions of financial intermediation, such as payments, risk management, savings and investment, there has also been the formation of OB due to copying. Connectivity has fostered imitation to economize on search costs and to benefit from shared networks. Copying can have adverse consequences, as when all economic agents congregate at the same hub, and, by adopting the same behavior, precipitate disastrous consequences as we saw in the 2008 financial crisis.

Keywords Information cascades · Network effects · Financial markets · Spontaneous feedback effects

The social and psychological forces that propel group formation in societies, such as homophily or the tendency to gravitate toward people with similar characteristics, can work in reverse, as with social influence or the

© The Author(s) 2017   105
S. Bhatt, *How Digital Communication Technology Shapes Markets*,
Palgrave Advances in the Economics of Innovation and Technology,
DOI 10.1007/978-3-319-47250-8_6

desire to remake oneself in the image of others. The Internet has spawned connections, some by deliberate choice but some created by the structure and behavior of the network. In any given connection, similarities between individuals will be understood but differences will be puzzling, at first. Repeated interaction will then move these individuals to erase these differences, as one or both copy characteristics and behaviors of each other. This phenomenon of imitation will then be replicated across the entire network in various strengths leading to systemic changes in the entire network. Connectivity not only brings information from node to node but it also makes their economic and social identities gravitate toward each other.

This phenomenon of copying is prevalent in all industries, but in the financial industry it can have massive consequences. "Finance is more like the circulatory system of the economic body" writes Alan Blinder [90]. Because finance is an important factor in all businesses, events in the financial industry will have ripple effects across the economy, enabling financial institutions to acquire the behemoth status. *Too big to fail institutions* are more than OB – these institutions are not simply large, their very survival is practically guaranteed. Congress offers these institutions a safety net so that they face limited probability of being dissolved. Dissolving an institution in this industry has far reaching consequences. In 2008, "Bear Stearns was deemed to be too *interconnected* to fail" [90].[1]

Paradoxically, both granularity and behemoths co-exist in the financial industry. The simple reason is that DCT has generated a cobweb of connections. In this chapter I examine, first, how increasing embeddedness, in the context of financial markets, dismantles traditional organizational structures, leading to granularity. When the decision makers are not the ones with the requisite information and resources, value maximizing choices may not be made. Granularity narrows the gap between these two groups, mitigating the agency problem. Second, increasing embeddedness connects seemingly isolated transactions across OB. When economic agents imitate choices made by others, the overall outcome is likely to be value decreasing, and quite possibly disastrous. As digital connectivity increases, the effects can sweep across the entire economy, well beyond financial markets themselves.

AGENCY AND GRANULARITY

Let me begin with the first problem. In a sparsely linked network, principals have to delegate decision making to other nodes, their agents, as well as share information with other nodes. Financial organizations are networks with central elements of authority, information, coordination, trust, responsibilities and rights, and processes for decision making. While these are standard elements of most organizations, financial networks have the unique feature of being intimately linked with other networks in the economic system and therefore have a comprehensive reach. A lapse in decision making or abrogation of responsibilities has repercussions beyond the organization itself.

This principal-agent problem is not new but increased connectivity has mitigated the problem by eliminating intermediaries, as we saw in Chap. 3. Formerly, an investor would select the fund manager, who selects firms, whose board members choose managers. At each link in this chain of transactions, there is the possibility of misallocation of resources. The investor, as principal, delegates management of retirement funds to the fund manager, as his agent. The contract between the two requires detailed information about the precise objectives of the investor and actions to take under various contingencies, such as a precipitous fall in the S&P500 Index. Implementation of this contract will require real-time coordination, stock market updates and a credible commitment to the specifications of the contract. Now imagine that this alignment of incentives and actions is required at the firm management level and then at the divisional level, and one can see the possibility of resources being depleted simply to monitor the process and set up an appropriate incentive structure.

With connectivity, new businesses leverage the tools of DCT to circumvent the inefficiencies and costs imposed by this agency problem – conflicts of interest between people who provide the resources and all the parties through which they travel. OR creates direct links between borrower and saver, and depositor and payments platform, thereby eliminating intermediaries. Granularity in the form of smaller companies is taking over individual business functions of financial institutions, along the lines of functionality: payments, agglomeration of resources, transfer of resources across time and space, risk management, and information revelation. "Fintech start-ups are nimble piranhas, each focusing on a small part of a bank's business model to attack" [91].

Granularity and Organizational Restructuring in Financial Markets

In financial markets, the functions of financial intermediation have been unbundled, which effectively increases granularity in the markets, reducing the agency problem. These functions can be classified as:

- Providing a medium of exchange (payments function),
- Store of value (investment function),

- Facilitating the transfer of resources across time and space (savings function),
- Risk management (hedging function) and
- Price discovery (liquidity function) in financial markets.

The Payments Function

The world of payments is changing as people are buying more goods online and increasingly with their phones. For example, security is considered tightest and convenience enormous in the NFC tap-and-pay technology (or near-field-communication technology). Used by Google Wallet and Apple Pay, payments are processed wirelessly by tapping the NFC-embedded device to an NFC terminal. In the case of Pay, the transaction is finished after a biometric fingerprint is provided. Credit card information is stored as an encrypted Device Account (DA) number on a secure chip inside the device. (It is stored in the Cloud with Google Wallet.) Even Apple does not have knowledge of this DA number or the individual transactions. Other apps, not based on the tap-and-pay technology, such as Bank of America's Mobile Pay app, allows the smartphone to be used as a point-of-sale device, which processes credit card transactions. PayPal uses the punch-in-identifiers technology which requires both hands to enter information into the app. Venmo, owned by PayPal, is a free app that allows users to transfer money to other users; merchants have only recently been added to this network. Both Square and Stripe, digital payment firms, have agreed to work with Apple to enable small businesses to accept Apple Pay. The transactional part of the payment is commoditized so their customer relation aspect, such as ease of use, differentiates the payments mechanisms.

These payments innovations have not eliminated bank deposits. They still require users to make the final payment using a bank account so banks as financial intermediaries will not be replaced. What may be eliminated is cash as a form of payment and while check use is on the decline, all new payment technologies ultimately link back to the bank deposit.[2] Significantly, by making processing easier, most new payments services are eliminating banks from the lucrative part of the payment supply chain – connections with customers are weakened. Banks are left to simply balance the ledger, a process that is likely to be automated by adapting blockchain, the technology behind the Bitcoin platform. As a result, banks will be receiving less information from transactions processed by them and therefore, less revenue generated by reselling this information to advertisers. For example, blockchain is used by R3, a company that is developing a blockchain network for use by the financial services industry. Bank of America, Goldman, HSBC and JPMorgan are now backing the company as they look to blockchain as a means of upgrading manual back office machinery in order to speed up asset transfers, lower operational and counterparty risk, and hence lower capital requirements [91].

With the advent of privately issued digital currencies, such as bitcoin, we have an unbundling of the services of a currency. For example, Coinbase, a San Francisco startup, which raised $75 million from investors such as Andreessen Horowitz in 2015, is at the vanguard of companies building businesses around bitcoin [91]. This firm has created an online wallet that allows people to store, send and accept bitcoin payments and also launched an exchange that allows individuals to change real-world money into the cryptocurrency.

The payments function in bitcoin is detached from the savings, investment, risk management and price discovery or liquidity function. Bitcoin serves as a payments mechanism so long as there is a party who will be willing to accept it as payment for goods or services. It is also a means of transferring resources across space: for example, a person in China can use bitcoins as payment for goods purchased in Estonia without transacting via a bank or financial intermediary. Bitcoin is a platform with strong network effects: as more people accept it, more people will become part of the network of users. Coordination, and common knowledge, among users suggests that the digital currency market could coalesce around a single currency, such as bitcoin.[3]

However, the volatility in the dollar-bitcoin exchange rate makes its role as a savings and investment vehicle, hedging vehicle and liquidity provider much less compelling. The absence of a global supervisory authority that would determine total bitcoin volume and maintain a stable exchange rate with respect to the dollar and other currencies remains a major entry barrier for bitcoin as a currency. As Rainer Bohme et al write, "Thus, bitcoin today resembles more of a payment platform than what economists consider a currency" [93]. But, they go on to conclude, "most users (by volume) treat their bitcoin investments as speculative assets rather than as means of payment."

Mobile payments in developing countries are providing an alternative to credit and debit cards. Mobile payments transfer money using wireless or newer technologies, but they require four features: mobile phone ownership; awareness of mobile payments; mobile infrastructure and a forward looking regulatory environment. There is limited mobile payment awareness in developing nations, but M-Pesa, a mobile phone app in Kenya, functions both as a temporary store of value and as a payments mechanism. Nearly 90 % of Kenyan households reported using mobile money services as of August 2014 and 56 % make or receive payments using cell phone. Mobile wallets have become a means of providing a stored value account through which payment can be received on a mobile device, and then turned into cash through an agent. Mobile accounts are more prevalent than bank accounts in more than fifteen countries, and, with more features and more interactivity, it becomes mobile banking [94].

In the US, the rollout of mobile payment technology was awkwardly implemented. Merchants were brought onto the platform before courting customers, hampering the driving force of network effects and thereby slowed the development of a large user base. Each side is waiting for the other to attain critical mass. Consumers are waiting for the ubiquity of a single payment method while merchants are waiting for sufficient network size. It may be efficient for other stakeholders in the mobile payment ecosystem to convince both consumer and merchants to use either the same technology or the technologies to become seamlessly compatible, as in the case of the hugely successful MobilePay in Denmark [94].

The Savings and Investment Function

Commercial banks recycle deposits by agglomerating them into larger loan contracts. Unbundling these two functions, crowd funding substitutes for the investment and saving function of banks, and also perhaps the risk management function. P2P lenders are websites that directly match borrowers with savers, often in specific sectors such as residential or commercial real estate (CrowdStreet Inc and Sharestates LLC) and movie productions (Junction Investments Inc) [95].

Risk Management Function

Pooling of risk across the network will enable large established insurance firms to survive and provide standard risk management contracts. The larger the risk pool, the greater the diversification across different risk categories so the negative outcomes associated with individual events are mitigated. Risk itself is not eliminated.

With the trajectory of labor markets moving in the direction of independent workers and independent contractors, as opposed to salaried employees, risk sharing contracts are likely to be replaced by more granular hedging arrangements. There is a market here for the introduction of job and disability related insurance arrangements and retirement savings plans. We are already witnessing granularity in health insurance under the Patient Protection and Affordable Care Act (ACA), which need not be purchased under the aegis of an employer.

Liquidity Function

Traditional financial intermediaries, such as banks, may be forced to reinvent their function in light of these digitally enabled developments. A single, well-known digital platform, such as Bitcoin, could overcome the coordination problem and not require compatibility between multiple digital currencies. During times of economic stability such a disintermediation could be sustained. However, an economy shocked by an external event cannot be resuscitated without government intervention (as we saw in the Bear Stearns case). The mere possibility of such an event, I believe, threatens the viability of a digital platform.

CONNECTIVITY, ORGANIZATION BEHEMOTHS AND SYSTEMIC RISK

Connectivity in financial markets is creating OB and providing the seeds via the imitation process for systemic risk or catastrophes. The same forces that I discussed earlier in Chap. 3, network effects and information cascades, are at play in the financial environment. Financial institutions being behemoths are not the problem – it is that they are too similar.

The problem of imitation arises where each node influences, and in turn is influenced by, the decisions made by other nodes – in other words, imitation or copying. Connectivity due to the Internet drives information sharing and cooperation, which then extends to collaboration and copying. Copying is to be interpreted broadly in that one node's behavior may form the template for another node's actions. The underlying logic is the fear of missing out (FOMO) from a rewarding outcome by not adopting a certain behavior.

Among the multiple causes of the potential "global financial meltdown" in 2008–2009, Alan Blinder lists three that have relevance here: the housing bubble or inflated asset prices, excessive leverage throughout the financial system and novel mortgage lending practices [90]. All of these had one factor in common: copying.

Bubbles, according to Blinder, are "a large and long-lasting deviation of the price of an asset from its fundamental value" [90].[4] The belief that house prices would continue to rise was based on observing and copying the actions of previous home buyers, who had accrued financial rewards. Blinder extends this argument as follows:

> You allegedly couldn't lose by investing in houses – which would rise in value by 10 percent or so a year forever... So too many American took on too much debt to buy houses they couldn't afford – and then refinanced them several times to pocket capital gains. The media fed the beast by hyping the good real estate news. [90]

There can be a tremendous impact of the network on social and economic processes by influencing decision-making. First, feedback effects, with individuals copying each other, can arise spontaneously or by design. Second, copying can arise from the benefits, or network effects, of joining a group. When the network shares a common product or characteristic or idea, network effects can also explain the diffusion of innovations and ideas. And third, the most extreme form of copying leads to power law effects, where all nodes congregate at some hub, adopting the same economic responses leading to catastrophic situations.

Spontaneous Feedback Effects

When individuals make decisions sequentially, and each individual has imperfect or fuzzy information, it is possible that decisions made by individuals earlier in the sequence will impact those making decisions later, since the latter group will ignore imperfect private information, or revise their private information, and instead follow the early decision makers. This leads to information cascades. The important assumptions here are (1) sequential decision-making where subsequent individuals observe the action but not the private information and (2) private information is imperfect or fuzzy and subject to some misinterpretation.[5] The rapid proliferation of novel financial instruments, such as collateralized debt obligations (CDOs) and credit default swaps (CDSs), during the financial crisis of 2008 is an example of this phenomenon.

Deliberate Feedback Due to Ranking

Firms can deliberately generate feedback effects by providing critical information. For example, *The Washington Post* provides an updated list of most emailed articles, inducing other readers to emulate and thereby increasing the likelihood of a future reader going to these articles. Further, if you were told how many *times* each article was emailed, this copying effect is likely to be exacerbated since the popularity differential between articles becomes public information. Knowing that article #1 on the list was emailed twice as many times as #2 will send even more traffic to article #1. If, on the other hand, the numerical difference between #1 and #2 is less than 1 %, traffic to both might be equal. Examples of this kind of deliberate feedback prevail in many media markets such as Netflix, with stars to denote rank. In financial markets, this could lead to panic and systemic risk if the *Post* article suggested some financial instability.

Are rankings themselves subject to some sort of information cascade? Users may start out with their own private rankings based on fuzzy, private information, but then upon observation of others' behavior, will update their own rankings such that ranking cascades are likely to form. For example, if all prior ranks of a local bank are 1 (on a scale of 1–5, with 5 denoting safety) then I am likely to ignore my own private information of 5 (the bank is safe) thinking that perhaps I was mistaken and withdraw my funds.

There are two significant assumptions being made in this analysis. First, the underlying assumption in a copying model is that all users are identical and trust-worthy. If reliability becomes a factor, then the copying model breaks down since each user's decision has to be weighed by their trust index. Second, we assume that an insignificant amount of time has lapsed between previous decisions and current ones. If all past users made their decision a long time ago, the product itself might have changed quality so current users may ignore past decisions and act as if they were the initial decision makers. Of course, this may start a fresh cascade in current time.

Network Effects

Second, another reason to follow other nodes comes from network effects, which can be *direct benefits* or *indirect benefits*. As explained in Chap. 3, direct benefits occur when more people join a network, increasing the value of joining to subsequent users. If more people use the payment mechanism Venmo, then the direct benefit of participating increases since you have a wider network of people to access. Indirect benefits are due to the benefits of coordination. If all users are on the same system, then there will be more complementary products produced, increasing the total benefit of the system. An example of this is the operating system, iOS. When more people use the iPhone or iPad, the more apps there are, increasing the value of the device.

Diffusion of an idea or financial innovation in networks works along the same principle as network effects. The benefits of synchronization increase not only as more members of a network adapt to the new innovation, but as more of a given individual's immediate neighbors adopt this new innovation. This idea is subtly different from network effects described above, where the fraction of the entire population that adopts a product is critical. Here, the threshold depends upon an individual node's *immediate* neighbors. If more neighbors adopt an innovation or idea, then the likelihood of adoption increases.[6] The switching threshold is reached when the fraction of trendy individuals exceeds the relative payoff from sticking to the old model.[7]

One can formulate examples, say a new mobile payment platform, where the sweeping diffusion of an innovation, called a complete cascade, depends upon the identity of the initial adopters and the network structure. A cascade will fail to sweep the entire network if there are dense clusters of individuals outside the set of early adopters. More strongly, we can say that a complete cascade will fail to take hold if and only if the the relative payoff from the new product is smaller than the fraction of non-adopting clusters [14].

Tightly knit communities, or clusters, can impede a cascade. Innovations from outside this tight community will find it difficult to find a way into this community. In Fig. 1.1(d), the local bridge G-A, linking networks with high clustering coefficients, may be a powerful way of conveying new information, but it also impedes cascades or diffusion of a risky new product. A bridge between networks only connects single nodes so its existence is not a compelling argument for switching products. Since the risk of adoption is mitigated if a high threshold of neighbors also adopts this innovation, risky new products and ideas are less likely to be transferred across local bridges.

An important example of impediments to the diffusion of ideas is the failure of collective action. Collective action is not simply a consequence of coordination problems, which requires access to localized and dispersed information. It also requires common knowledge, which is a higher order of information access. It is the assurance that all participants in the network have access to the very same body of information. For example, if a coordinated response is necessary to abandon a particular financial innovation or fire a CEO, then a critical minimum (threshold) number of neighbors or colleagues need to be in mutual agreement as to the action and the desired outcome. This action relies on the firm and secure knowledge that not only are other members acquiescent with the plan, but others know that they know and so on. Importantly, each individual is going to use immediate neighbors' threshold as a signal for the participation of the entire network. And here is where the tricky part comes in. Suppose all residents of a village are located on a straight road so each resident has two neighbors. Also suppose that there is a plan to fire the village headman but that each resident will participate only if three other residents participate. If villagers know only the views of their immediate neighbors but not that of other members of the village, the idea may never get off the ground. This can happen even when most members of the village would otherwise have accepted the idea. The problem is one of common knowledge, not just coordination.

Rich-Get-Richer or Power Laws

Third, an empirical fact that corroborates this notion of copying is the existence of the Power Law or the rich-get-richer phenomenon. For example, the fraction of webpages with exactly k in-links is monotonically decreasing in a non-linear fashion and is proportional to $\frac{1}{k^2}$.[8]

In networks, the Power Law best describes the formation of hubs, or nodes with many in-links. The dynamics of hub formation has three stages. First, there is growth of the network, as new nodes are being added over time. Second, these nodes link to senior ones via an algorithm called preferential attachment. Basically, a new node links to hubs or nodes with a large number of in-links. The more in-links a senior node has, the more nodes will preferentially link to this node. This dynamic is called the "rich-get-richer" rule, which predicts that senior nodes will become hubs conferring some sort of a first-mover advantage. In some markets, however, we do observe latecomers frequently outperforming first-movers. This leapfrogging of newcomers may be due to the quality of the individual nodes, as explained below.

Third, there is the notion of fitness of nodes. The assumption that all nodes are equal is clearly not valid. This feature is captured by "fitness," defined as "a quantitative measure of a node's ability to stay in front of the competition." Then "a simple way to incorporate fitness into the scale-free model is to assume that preferential attachment is driven by the product of the node's fitness and the number of links it has" [96]. Fitness captures the notion of speed such that a node with high fitness grabs links faster than nodes with the same seniority but lower fitness. For example, Google, whose search engine is superior, acquired users much faster than Yahoo.

The phenomenon of hubs has implications for the survival of the Internet. The architecture of Internet is such that for the entire network to collapse, multiple interconnected hubs must be destroyed simultaneously. If the probability of a node being attacked or otherwise rendered dysfunctional is equal to p and is the same, but independent, across all nodes, then the probability of multiple nodes being destroyed simultaneously can be shown to be small. For example if there are n critical hubs, then for the entire network to crash we need all n nodes to break apart – the probability of this occurrence is p^n, which tends to zero as n goes to infinity since $0 \leq p \leq 1$. The Achilles' heel is the appearance of an informed hacker who selectively and systematically targets critical hubs and brings the entire network to a halt.

My Take

Connectivity breeds copying with possibly adverse consequences. Observing the law or following social norms are also forms of copying, but are a consensual aspect of social behavior. What we have discussed in this chapter is a replication of others' behavior based on limited information, on the part of the originator of the action as well as followers. Systemic risk in financial markets arises from precisely this form of copying behavior. Copying without a clear individual agenda is the issue here.

When Alice asked the Cheshire Cat for directions, she was told "That depends a good deal on where you want to get to." Upon answering that she didn't care, the Cheshire Cat replied, "Then it doesn't matter which way you go" [97].

Granularity in products and organizations is the force that could inject personalized trajectories, mitigating copying due to the FOMO.

NOTES

1. The fifth largest Wall Street investment bank, Bear Stearns was on life support over the March 15–16, 2008 weekend and remained solvent on Monday, March 17 only after a $30 billion cash infusion from the Federal Reserve via the regulated commercial bank, JP Morgan Chase [90].

2. The desire for anonymity, safety and insurance will remain the bedrock for a significant demographic, which will use and keep a cash reserve. The worldwide circulation of the most popular $100 billion increased 348 % over the past twenty years; the next most popular $20 billion increased 103.4 % over the same time period [92].

3. Following convention, bitcoin, the currency has a lower case b, and Bitcoin, the technology platform has an upper case B.

4. Bubbles in speculative markets are founded on the imitation phenomenon. Blinder concludes that "the herding behavior that produces them may well be programmed into our DNA" [90].

5. The theoretical explanation of information cascades is Bayes' Rule. If $P(A)$ is the prior probability of event A and $P(B/A)$ is the conditional probability of event B given A, then Bayes' Rule says that $p(A/B) = \{P(A)/P(B)\}*P(B/A)$. $P(A/B)$ can be interpreted as the posterior probability of A, now that B has occurred.

6. This discussion of diffusion of ideas is based on Easley and Kleinberg [12a]. More formally, the threshold for switching to a new product, p, depends upon the relative payoffs. Let p and $(1-p)$ be the fraction of an individual's neighbors adopting the new product A and the old product B, respectively. Suppose the person has x neighbors and that the payoff from adopting the new product is a, while the old product has payoff b. Then switching to the new product is the best choice if:

$$pxa \geq (1 - p)xb$$
$$\text{or } p \geq \frac{b}{a + b}$$

7. In the marketing world, disseminating new product information can be structured after this threshold model. Production introduction can be diffused more widely if the threshold is lowered (payoffs raised by raising quality) or if key nodes are *infected* with the product, either with lower prices or some promotion. Chapter 5 discusses the role of gatekeepers in infecting key nodes for advertising purposes.

8. It has been found in [96] that this fraction follows a Power Law

$$f(k) = ak^{-c} \text{ or } \log f(k) = \log a - c \log k.$$

The unique feature of the Power Law distribution is that there is no characteristic node or average number of links for nodes, nor a well-defined standard deviation, as is common in the bell curve or normal distribution.

BIBLIOGRAPHY

[90] Blinder, Alan. *After The Music Stopped*. New York: The Penguin Press, 2013.

[91] "Developed World Plays Waiting Game with Mobile Payments," *Financial Times*, October 16, 2015.

[92] Author's calculations based on data from the Board of Governors of the Federal Reserve System, retrieved May 2, 2016 from https://www.federal reserve.gov/paymentsystems/coin_currcircvalue.htm

[93] Bohme, Rainer, Nicolas Christin, Benjamin Edelman, and Tyler Moore. "Bitcoin" Economics, Technology, and Governance." *Journal of Economic Perspectives*, Spring, 12–15, 2015.

[94] Bhatt, Swati, "Why Don't More Americans Use Mobile Payments," *International Banker*, Spring 2016

[95] Parmar, Neil, "How Crowdfunding Opens Doors Long Closed to Most Investors," *Wall Street Journal*, November 10, 2014

[96] Barabasi, Albert-Laszlo. *Linked: The New Science of Networks*. Cambridge, MA: Perseus Publishing, 2010.

[97] Carroll, Lewis. *Alice's Adventures in Wonderland*. London: Macmillan and Co, 1865.

CHAPTER 7

The Private World of Sharing and Cooperation

We build lines in the sand, not walls, so we can see, hear and step across

Abstract Connectivity builds bridges between individuals' private space and threatens notions of identity and privacy. While the notion of privacy as anonymity is relatively new from a historical point of view, the constitutional guarantees pertain to lines of control over what we share, when we share and with whom we share. Anonymity implies the absence of any form of identity so the requirement is not one of erasing identity but rather preserving that identity in a secure and undisturbed form. People don't want to vanish into obscurity, they want to be known as obscure, private individuals. There is, however, the reality that DCT has swamped our ability to create privacy shields and that transparency may be the new normal.

Keywords Freedom speech · Choice of medium and message · Anonymity versus privacy · National security

Major technology policy issues fall under the First Amendment, freedom of speech, and the Fourth Amendment, the right to privacy. The First Amendment states:

S. Bhatt, *How Digital Communication Technology Shapes Markets,*
Palgrave Advances in the Economics of Innovation and Technology,
DOI 10.1007/978-3-319-47250-8_7

> Congress shall make no law respecting an establishment of religion, or prohibiting the free exercise thereof; or abridging the freedom of speech, or of the press; or the right of the people peaceably to assemble, and to petition the Government for a redress of grievances.

The Fourth Amendment states:

> The right of the people to be secure in their persons, houses, papers, and effects, against unreasonable searches and seizures, shall not be violated, and no Warrants shall issue, but upon probable cause, supported by Oath or affirmation, and particularly describing the place to be searched, and the persons or things to be seized.

In a heterogeneous society, are there any bounds of ethics, morality and law to free speech so that inflammatory words are circumscribed? What constitutes "search and seizure" and under what "probable cause"? In this chapter, I first consider how DCT is stretching, and perhaps changing, the substance behind these amendments. Then, the central questions of privacy, anonymity and identity are examined. Are there institutions, both business and state, capable of monitoring these new values to ensure the integrity of the law?

FIRST AMENDMENT

Limits to the freedom of speech are already being defined by private interests. While recent data suggest that young people, aged 18–29, feel that the state should intervene in preventing speech that could be regarded as offensive to minority groups, limits to the freedom of speech are already being defined by private interests in the US.[1] On university campuses across America, as students are drawing attention to past injustices, they are simultaneously acting as censors of free speech by preventing talks by Condoleezza Rice, Ayaan Hirsi Ali (a Somali writer and Muslim) and Jason Riley (an African American author). "Students and their sympathizers think that free speech is sometimes invoked to deflect these claims; or, so Princeton's Black Justice League maintains, as a 'justification for the marginalization of others' " [99].

Recently, a gossip blog, Gawker.com, declared bankruptcy after a $140 million lawsuit alleging violations of privacy. The lawsuit was financially supported by Peter Theil, the tech billionaire, who writes

A free press is vital for public debate. Since sensitive information can sometimes be publicly relevant, exercising judgment is always part of the journalist's profession... The press it too important to let its role be undermined by those who would search for clicks at the cost of the profession's reputation. [98]

When Facebook and other social media platforms set limits on content publishable on their sites, they are being private enforcers of the First Amendment. However, when Apple opposed inclusion of the rifle emoji as part of Unicode 9, the standard set by the non-profit consortium Unicode, the Harvard scholar, Jonathan Zittrain questions the extent of private sector vigilantism.

To eliminate an elemental concept from a language's vocabulary is to reflect a sweeping view of how availability of language can control behavior, as well as a strange desire for companies – and inevitably, governments – to police our behavior through that language. In the United States, this confuses taking a particular position on the Second Amendment, concerning the right to bear arms, with the First, which guarantees freedom of speech, including speech about arms. [100]

FOURTH AMENDMENT

The Fourth Amendment protection expanded significantly with the *United States vs. Jones* (2012) case. Law enforcement officers in a drug-related investigation had installed a GPS device for a month on a car's exterior without Jones's knowledge or consent. This was not only a test of "unreasonable searches and seizures" without a warrant, but also government intrusion upon privacy and civil rights. The Court concluded that since the car was Jones' property, the intrusion on the vehicle was for the purpose of obtaining information and therefore a search under the Fourth Amendment, for which a warrant would have been needed [101].

More significantly, the Court's concurring opinion, by Justice Alito, was that a mosaic of information could be created from digital scraps of data which meant that the GPS device information, while minor, could generate a larger picture or mosaic of Jones, violating Fourth Amendment privacy rights. Alito argued that "relatively short-term monitoring of a person's movements on public streets accords with expectations of privacy that our society has recognized as reasonable." However, *extended* monitoring, which enabled observation of an individual's movement patterns, thereby generating a personal story, violated the Fourth Amendment [101]. This was a landmark case in that violation of Fourth Amendment was interpreted in terms of both privacy and personal property.

The European Union Data Protection Directive goes further in making privacy a fundamental right. This includes the right to be forgotten so that individuals can request search engines to remove links that are no longer relevant. This feature suggests some form of editorial control by citizens over what might be public information and also places the burden of removal on the search engine, the platform, rather than the original information source. Additionally, this information is not removed from all global websites.[2]

The Anonymity–Identity Spectrum

At the heart of both Amendments is the idea of connectivity. First, we have the freedom to choose both the medium and message of communication. What determines how we communicate and how do most people want to communicate? We have the option of communicating via direct face-to-face contact or via voice (phone), text, email and photo images. The available communication technology determines the scope and range of our words. Digital technology generates viral connectivity so First Amendment rights have far more power than originally imagined. Second, technology also determines the extent to which this communication remains between the interested citizens – this choice, therefore, impacts privacy outcomes. Consider that the very technology that enables our communication is created by business: as a consequence, the boundaries of our private lives are defined by the technological choices we make. Most individuals remain unaware of the sieve-like nature of transmission technology and how it can be penetrated at various points by businesses for monetary gain and by the state for national security purposes. Importantly, neither does the law! The Privacy Act of 1974 is based on a set of principles that are not synchronized with current technological capabilities.

From an economic perspective, what does freedom of speech imply? Magazines sell and advertisers pay for space on platforms that host popular content, and when popularity is reinforced by titillation and voyeurism, there is an incentive to seek and publish intimate details about individuals. This is where the First and Fourth Amendments collide. In free and unconstrained markets, businesses have the right to pursue their objective of profit maximization (which could be defined to encompass all stakeholders in the business), but this right is incompatible with individual rights to their private information. The significant question for the courts and for civil society is how should violations of civil liberties, embodied in both the First and Fourth Amendments, be addressed?

What do we mean by privacy? It encompasses several definitions. One is a form of autonomy or independence of personal choice, including marriage, religion, secrecy or hiding personal conduct, and anonymity or the ability to conduct your life totally unobserved. In many nations, including the US, autonomy in personal choice is guaranteed by law. But, as discussed in the case of the market for individuals' attention and labor markets, the central issue is one of control. If privacy is considered a basic human right, with ownership status as with property rights, then individuals have the right of control. Control over whether, when and with whom to share their private information. We have largely accommodated this notion with most economic transactions, which are paired with complex privacy agreements. (It's not clear if those who profess to believe in privacy rights actually read these agreements prior to signing off.)

The second form of the privacy definition addresses the issue of government surveillance while the third form covers the notion of privacy as a basic human right. These two definitions expose a spectrum of private information where there is complete anonymity on one end and total transparency on the other.

Anonymity also implies the absence of any form of identity. A world where individuals are totally unknown has never existed. Membership in tribes, kin groups and communities automatically generates an identity. In fact, as discussed in Chap. 3, socialization is ingrained in human nature so that we may acquire an identity. The fear of being alone, with no kith or kinship ties, is precisely what drives civilization and culture. The idea of complete anonymity is antithetical to any evolutionary trajectory so when we argue about privacy, we are asking a narrow question. The question is not one of absence of identity but rather preserving that identity in a secure and undisturbed form. People don't want to vanish into obscurity, they want to be known as obscure, private individuals.

Privacy and Business

Consumers have already made a Faustian bargain with business when conducting any non-cash transaction. Digital transactions and payments leave retraceable trails so any and all data imparted is in the public sphere. There is directionality to this economic connection, since once information is sent from buyer to seller, it cannot be retrieved or retroactively erased. In many instances, more information than strictly necessary is imparted in exchange for price discounts for merchandise (or freebies).

Most of us divulge personal data routinely on social media, search engines and shopping web sites, allowing third parties to commercialize this data. In addition to voluntary sharing of information, social media enable (even encourage by providing the tools) surreptitious private surveillance which is more commonly known as stalking. This process uncovers data that are already "out there" – it is often simply a matter of connecting the dots and uncovering the pattern of an individual's life. We litter our public space with intimate details of our lives, in tiny crumpled pieces of data that we think are only worthy of disposal. Just as governments pore over the trash of potential terrorists, these pieces of data can be stringed together in a private surveillance effort. While there are rights to privacy, there are also freedom of speech rights, so does each individual have the responsibility to watch what he throws away or does the voyeur have the responsibility to exercise restraint? I believe that just as you are a visible public figure whenever you enter public spaces, like walking on a street, your personal information is public as soon as you digitally connect with the world. Gossip, in the pre-binary days, was based simply on hear-say and not hard evidence; in the digital world, we inadvertently leave a trail of evidence.

A recent survey by the Pew Research Center found that while the notion of privacy on the Internet means different things to different demographic groups, the overall concern is that their personal information is no longer secure. Eighty percent of individuals who use social networking sites feel insecure about the use that advertisers might make of the information shared on these sites. Furthermore, 80 % of adults agree that the public should be concerned about government collecting data via phone and Internet communications. Paradoxically, 55 % of those surveyed would share information in order to obtain some online services for free.

Landlines were considered the most secure form of communication; 67 % of adults surveyed felt secure or very secure about landlines compared with 39 % feeling secure about text messages and 40 % about email [102].

Think of the information market as consisting of suppliers (consumers who reveal preferences through search and transactions), intermediaries who collect, curate and store this information and final users or firms that buy this aggregated data to improve products and prices. Mann [103] makes the case that there are economies of scale and scope associated with information. The economies of scale arise due to the increasing value of information aggregated across time for a single individual. To the extent that an individual's behavior is correlated across time, knowing a customer's history allows the firm to generate a finer picture of his preferences. Economies of scope are generated when information is aggregated across wide swaths of consumers, allowing firms to tailor products to specific groups, and also price discriminate more efficiently, by charging different prices to groups with similar characteristics.

Thus, customers face a tradeoff between hiding and sharing data. For example, it is more convenient to share personal information with a bank in order to facilitate online banking. But it may be costly (in terms of loss of privacy) to share this same information with a retailer who might himself face a tradeoff between guarding this information or selling it to third parties. Further, consumers may face bounded rationality in that they may undervalue the cost of revealing information. Innocently providing phone numbers and email addresses at point of sale, which is unnecessary for most transactions, allows the business to initiate communication with the individual precipitating a string of interactions. From a policy perspective, how should this tradeoff be balanced? As Mann writes,

> Policy makers and businesses as well differ in their response to the limited rationality of consumers. The EU Privacy Directive is at one extreme, disallowing the collection and retention of personal information on the grounds that consumers don't know what they are giving up. Other policy approaches require active consent (opt-in) or more transparency ('this website uses cookies. . . . click here for our cookie policy'). [103]

Malevolent use of customer information is a larger concern than that of information use by third parties for commercial reasons. There is the more important problem, and one that is occurring more frequently, of data breaches. If valuable financial information is stolen from a credit card database, it could compromise multiple consumers. Disclosure of data breaches is necessary not only for transparency but also for calculating costs and benefits of providing and protecting information, and of assigning responsibility and costs in the case of a data breach.

The target data breach in the fall of 2013 focused on credit card transactions. The company passed on some of the blame on credit card companies for failing to use the chip-and-pin technology (widely used in Europe), which would have minimized the likelihood of this data breach.

PRIVACY AND NATIONAL SECURITY

In the aftermath of 9/11, the President's Surveillance Program, authorized on March 10, 2004 by the White House Counsel (not the Department of Justice), allowed the government to gather metadata from Internet Service Providers without the consent of the parties involved.[3] President Obama signed an extension to certain provisions of the program on May 26, 2011 which allowed for foreign surveillance of suspected terrorists and access to certain information, metadata, from ISPs.

The revelations offered by Edward Snowden "fall into three categories: how the NSA uses the Internet and cellular networks to spy on non-Americans, how it uses the Internet and cellular networks to engage in economic espionage and how it uses the Internet and cellular networks to spy on Americans" [104]. The National Security Agency program called PRISM "allowed the NSA to obtain virtually anything it wanted from the Internet companies that hundreds of millions of people around the world now use as their primary means to communicate" [104]. This allowed the NSA access to private documents – emails, text messages, phone calls and documents in the Cloud – without a warrant, perhaps violating the Fourth Amendment.

Defense of the state necessitates surveillance activities, but in a democracy these have to be conducted within the bounds of the law. Citizens have conflicting demands in maintaining their anonymity and in securing their state. What point along the spectrum of anonymity–transparency is society going to choose? As citizens, we have the right to national defense against external threats. In pursuing this objective, the state may require access to private information so we have "delegated transparency" to select state institutions [105].

ENCRYPTION

Encryption can be used to ensure that communication between two parties is protected or to protect data at rest, that is, data residing on a hard drive. Many communications and technology firms may encrypt users' communications that pass through or reside on their servers, but the firms hold the key to decrypt the data. With end-to-end encryption, though, only the computers at each end of a communication have the encryption keys and only they can read the messages, with the service provider itself unable to read the data.

The FBI wants some sort of special knowledge of encrypted systems, which will give law enforcement access to a master encryption key (or keys) needed to decrypt data residing on or passing through a system. This special access creates a single point of vulnerability into systems, exactly what organizations and individuals are increasingly trying to avoid. However, companies feel that special access not only undermines the confidentiality of data, but also its authenticity (i.e., hackers who acquire the master keys would be able to forge communications and make them look legitimate).

At the crux of this debate is the fact that special access provided to law enforcement undermines the security of systems. If true, then this is a zero-sum situation: either cybersecurity is paramount or law enforcement gets special access to catch bad guys. That said, legitimate questions have been raised about just how absolute is this trade-off. For example, Apple has implemented a deliberate strategy of credible commitment to non-cooperation by encrypting devices such that it retains no decryption ability [106]. On the other hand, companies like Google have the ability to decrypt Gmail communications for precise advertisement targeting and there appears to be no security issues emanating from this business strategy. The security vulnerability resulting from special access is one of the fundamental points in this debate, and we need clarification for the basis of that vulnerability.

Even if Congress mandates some sort of special access, there is no guarantee that it would be the effective solution law enforcement wants, since criminals and terrorists could just buy different products that do not have a backdoor. Apple and Google may be required to provide special access, but what about a company outside US jurisdiction that sells end-to-end encryption communications applications? Is it possible to maintain both cybersecurity and special access?

SHARING OF INFORMATION ACROSS AGENCIES: DOES THIS LEAD TO EXCESSIVE TRANSPARENCY AND AN AUTHORITARIAN STATE?

When the FBI places an individual on a terrorism watch list, they have the right to any and all information pertaining to their investigation. Frequently, domestic criminal data obtained by the CIA is not shared with the federal intelligence community (FBI) due to the Foreign Intelligence Surveillance Act (FISA), which was enacted in 1978 to protect against the excesses of surveillance and potential invasions of privacy. The wall that was created between the FBI and the CIA created a compliance culture within the FBI and the Department of Justice, which houses the Office of Intelligence Policy Review (OIPR) [105]. OIPR became effectively the wall and the gatekeeper between the CIA and the FBI, preventing the sharing of information which, as later events confirmed, is vital.

A horrific recent example is the mass shooting in Orlando, Florida on June 12, 2016. The shooter was investigated by the FBI for ten months beginning in 2013 and placed on a terrorism watch list, but the probe was closed in March 2014. If the shooting was a case of "homegrown extremism" as President Obama called it, then this raises the question of sharing of information across agencies whose directive is to protect US citizens. The CIA is in charge of domestic surveillance and would have had knowledge of activities linking the shooter to prior shootings in the Boston Marathon case [107].

On the table currently (by 2017) is reauthorization of Section 702 of FISA. This Section targets non-US people (both citizens and permanent non-resident aliens) located outside the country for the purpose of intelligence collection. However, critics claim that in the process of collecting this information, US citizens' privacy rights might be violated, in a manner akin to the telephone metadata collection revealed by Edward Snowden. This piece of legislation is linked to President George W. Bush's Patriot Act, which many feel was an overreach of executive privilege. Compliance with this Act is required from traditional ISPs such as ATT, and also OTT platforms such as Google and Facebook. Under PRISM and "upstream collection" not only is information about the identity of individuals collected but also the content of their communication.

WHEN PRIVACY IS A PUBLIC GOOD

Shared private information creates a common resource, much like a public park, where individual contribution of private data adds up to a data pool that has vast social and community benefits. Just as taxes pay for public goods such as parks, shared data "pay" for common resources or public goods. Connections across the network economy make the benefits of sharing apparent so cooperation is elicited.

Large, detailed private data are more valuable when shared because they provide the information for improved transportation and public health systems. Common traffic patterns in congested cities and movement patterns in infected areas can be analyzed by data downloaded from personal databanks on smartphones. For example, in 2014 when Ebola raged in West Africa, doctors traced geolocation capabilities on mobile phones to contain spread of the virus. Orange Telecom in Senegal cooperated by giving the data to the Swedish non-profit Flowminder, which aided in drawing up population movement maps. Doctors then set up treatment areas and quarantine areas. Similarly, the US Centers for Disease Control is collecting activity data from mobile phone operators to view where most helpline calls are coming from, since an increase in mobile calls from a single area would alert authorities about a potential disease [108].

Drone technologies manifested in unmanned aircraft systems (UAS) are basically surveillance platforms that enhance observational capabilities. They are used for environmental monitoring of pollution in cities and tracking wildlife in National Parks, for example. They are also used by the media as a safe platform for news gathering and by package delivery services such as Amazon. However, there are also malevolent privacy invasion possibilities such as when an UAS flies over private property and collects personal data. The Federal Aviation Administration requires all UAS weighing between 0.55 and 55 lbs to be registered, which suggests an incremental approach to UAS regulation [109].

Private sites collect the vast majority of personal data and these data sets are more valuable as a common resource since their social value far exceeds their private value. For example, Nextdoor is a social networking site for neighborhoods, founded in 2011, and enables connections between neighbors for sharing community-related information. Nextdoor Now can help residents find local services and neighbors' reviews. The content on this site is crowd sourced with personal information, but to the benefit of all. Malevolent use, such as racial profiling, is an unforeseen outcome of the original intent of the neighborhood crime watch. "Rather than bridging gaps between neighbors, Nextdoor can become a forum for paranoid racialism" [110]. While there is the potential for any private crowd sourcing site to marginalize and exclude, Seattle police department practices "micro-community policing" by using local data to tailor law enforcement to the relevant community [111].

MY TAKE

The US Constitution was written at a time when connectivity meant horses and carriages. The transcontinental railroad had just entered the radar screen, but its implications were never imagined. Reinterpreting the law in the context of today's technological environment might be arduous and arbitrary. But it may be time to rethink our basic values and frame our analyses in a language of connections. Why do we want privacy, while simultaneously maintaining our right to freely trespass another's attention with our words?

NOTES

1. According to a recent article in the *Economist*, 21 % of German youth, 47 % of British youth and 55 % of French youth, aged 18–29, favor government intervention [98]. Reporting spam has become the new normal on social media sites such as Facebook. Is it easier to write hate messages or are young people just becoming more intolerant?
2. The Charter of Fundamental Rights of the European Union has two distinct conceptions of privacy. First, there is respect for private and family life and second, everyone has the right to the protection of personal data concerning themselves. The corollary to the second concept is the right to be forgotten.
3. Metadata is a higher form of data, covering not simply the identity of individuals under surveillance, but the content of their communication, including names of third parties, who may be unrelated to the mission on hand and whose privacy is therefore being violated. This issue is at the heart of the debate over surveillance activities, foreign and domestic.

Bibliography

[98] Thiel, Peter. The Online Privacy Debate Won't End With Gawker, *New York Times Op-Ed*, August 15, 2016. Accessed from http://www.nytimes.com/2016/08/16/opinion/peter-thiel-the-online-privacy-debate-wont-end-with-gawker.html?emc=eta1

[99] "Don't be so Offensive," *Economist*, June 4, 2016. Accessed on June 20, 2016 from http://www.economist.com/news/international/21699903-young-westerners-are-less-keen-their-parents-free-speech-dont-be-so-offensive?frsc=dg%7Ca

[100] Zittrain, Jonathan. "Apple's Emoji Gun Control," *New York Times Op-Ed*, August 16, 2016. Accessed from http://www.nytimes.com/2016/08/16/opinion/get-out-of-gun-control-apple.html?emc=eta1&_r=0

[101] Thompson, Richard. "*United States vs. Jones*: GPS Monitoring, Property and Privacy," *Congressional Research Service*, April 30, 2012. Accessed June 15, 2016 from https://www.fas.org/sgp/crs/misc/R42511.pdf

[102] Madden, Mary. "Public Perceptions of Privacy and Security in the Post-Snowden Era." http://www.pewinternet.org/2014/11/12/public-privacy-perceptions/. Accessed November 17, 2014

[103] Mann, Catherine. "*Information Lost*," 2013, NBER Working Paper 19526

[104] Halpern, Sue. "Review of No Place to Hide: Edward Snowden, the NSA, and the US Surveillance State by Glenn Greenwald" (2014), *New York Review of Books, University Press Issue*, July 10, 2014

[105] Rosenzwieg, Paul. *The Surveillance State: Big Data, Freedom and You*, The Great Courses Audiobook, 2016.

[106] Hennessey, Susan, and Benjamin Wittes. "Apple is Selling You an IPhone, Not Civil Liberties," Brookings, February 18, 2016. Accessed June 16, 2016 from http://www.brookings.edu/blogs/techtank/posts/2016/02/18-apple-iphone-civil-liberties-hennessey-wittes

[107] Wall, Matthew. "Ebola: Can Big Data Analytics Help Contain its Spread?" BBC, 10/15/2-14. Accessed June 14, 2016 from http://www.bbc.com/news/business-29617831

[108] Goldman, Adam, Matt Zapotsky, and Mark Berman. "FBI had closely scrutinized the Orlando shooter before dropping investigation," *Washington Post*. Accessed June 14, 2016 from https://www.washingtonpost.com/world/national-security/fbi-had-closely-scrutinized-the-orlando-shooter-before-dropping-investigation/2016/06/13/838e9054-3177-11e6-8ff7-7b6c1998b7a0_story.html?wpisrc=nl_headlines&wpmm=1].

[109] Federal Aviation Administration. "Unmanned Aircraft Systems." Assessed June 15, 2016 from http://www.faa.gov/uas/

[110] Harshaw, Pendarvis. "Nextdoor, the Social Network for Neighbors, is Becoming a Home for Racial Profiling," Fusion, March 24, 2015. Accessed June 15, 2016 from http://fusion.net/story/106341/next door-the-social-network-for-neighbors-is-becoming-a-home-for-racial-profiling/

[111] Waddell, Kaveh, "The Police Office 'Nextdoor," May 4, 2016. Accessed on June 15, 2016 from http://www.theatlantic.com/technology/archive/2016/05/nextdoor-social-network-police-seattle/481164/

CHAPTER 8

The Internet and Regulation

Internet freedom often necessitates oversight

Abstract The Internet was created in a culture of collaboration and distributed decision-making. With organization behemoths co-existing with granularity, has the Internet lost this collaborative culture? Ownership of data, of domain names and control over the flow of content in the entertainment industry (net neutrality) have become central issues in the digital economy. The concern with behemoths is not pricing power but rather the power to shape ideas by controlling content. Who curates and regulates global content with a view to fairness and balance? While social media are part of this larger picture the trajectory from cause to effect is ambiguous – there is no individual directing the show, only computer code. Virtual space is a shared resource whose free consumption necessitates governance of this commons.

Keywords Data brokers · Domain names · Net neutrality · Social media as gatekeepers

The Internet was created in a culture of collaboration and distributed decision making. Paul Baran published the idea in 1960 that "there should be no main hub that controlled all the switching and routing . . . control should be completely distributed" [23]. Information would be broken down into identical sized packets and transmitted across the network such that every node had equal power to channel the packets and that no central authority could

© The Author(s) 2017   133
S. Bhatt, *How Digital Communication Technology Shapes Markets*,
Palgrave Advances in the Economics of Innovation and Technology,
DOI 10.1007/978-3-319-47250-8_8

dominate the system. Well before that, the most fundamental discovery of the Internet was the transistor in 1949 and it was created from an open collaboration between two scientists in

> an environment where they could walk down a long corridor and bump into experts who could manipulate the impurities in germanium, or be in a study group populated by people who understood the quantum-mechanical explanations of surface states or sit in a cafeteria with engineers who knew all the tricks for transmitting phone signals over long distances. [23]

With OB co-existing with granularity, has the Internet lost this collaborative culture? Many of the scientists associated with the early development of the Internet were couched in the anti-war and anti-establishment culture of the 1960s. The culture was one where creativity and community were synchronized, culminating in the publication of the famous Whole Earth Catalog by counter-culture activist, and cyber culture promoter, Stewart Brand in 1968. The first network connection, the ARPANET, was made in October 29, 1968 between a computer at UCLA and one at the Stanford Research Institute (SRI) in a spirit of open source development of software.[1] Was the free, open and innovative Internet simply an isolated episode in the development of global communications [23]?

John Parry Barlow, former lyricist of the Grateful Dead, and co-founder of the Electronic Frontier Foundation, argued that for cyberspace, it was "not just that government *would* not regulate cyberspace – it was that government *could* not regulate cyberspace. Cyberspace was, by nature, unavoidably free" [18]. This notion of a libertarian state that would be self-ordered is very suggestive of the trend towards cooperation that we are witnessing today.

This chapter asks what core values of the Internet need protection and what are the appropriate institutions that can implement this? We will first consider ownership rights over the massive data collected in cyberspace. Virtual space is a shared resource so there is the problem of governance of the commons, of digital space. Who defines and enforces the rules that govern the space of digital technology? The data that we create by interacting in digital space is public property, so who has governance rights over this BD? This question is separate from that of privacy, particularly in the binary world, since we may have anonymity until we interact, but once we conduct business we have exposed that transaction to all forms of data collection.

Related to the idea of data as public property, is the question of location and address in cyberspace. While the early history of the Internet was written by two dynasties, hackers and university researchers, the assignment of virtual addresses today has entered the realm of business.

Finally, who regulates and curates global content with a view to fairness and balance? How do we address the concentration of economic power in the hands of OB – the big 10: Google, Amazon, Facebook, Apple, Netflix, Microsoft, Salesforce, eBay, Starbucks, Priceline? The more complex world of the Internet in 2016 is different from the simpler one of 1968 and self-regulation may be insufficient to address complex national security concerns. I focus on the issue of concentration of power and the narrowing of our choices that this could imply. Basically, do OB generate the benefits of a free market economy – innovation and choice? If we impose restrictive legislation are we impeding the incentive structure for startups or are we creating a more nurturing environment for new firms?

BIG DATA AND OWNERSHIP

The Federal Trade Commission has required that commercial data brokers be transparent and give consumers control over their personal data. Commercial data brokers (CDB) are private firms, including Google, that engage in collecting and aggregating vast publicly available information about an individual, such as addresses, telephone numbers, email addresses, demographic information and real estate data. CDBs typically sell their information for identity verification, marketing, and other purposes such as locating friends and nefarious stalking. More sophisticated brokers correlate disparate bits of information and, using pattern recognition software, generate a clear picture and history of individuals [112].

Every time individuals connect with the Internet they are giving away personal data. The plethora of possibilities made available by the Internet of Things (IoT) opens up multiple avenues for connecting with the Internet and divulging information. Nest, the IoT platform owned by Google, that monitors homes while the residents are away, has detailed knowledge about the movements and lifestyle of these absent individuals. Who owns this data? With ownership rights come the privileges of trading this data and the possibility of malevolent use.

The law applicable to protection of email communications, text messages, social media messages, videos and other metadata (non-content information such as date/timestamps, to/from information) is the 1986 Electronic Communications Privacy Act (ECPA). While this law applies to businesses generally, it does not apply to government access to these records since there are "procedures under which the government can require a provider to disclose customers' communications or records" [4]. Title II of this act is the Stored Communications Act (SCA) which covers access to stored emails directly from the service provider. The SCA applies both to email providers as well as data storage companies, and while voluntary disclosure of data is prohibited, it contains "procedures permitting the government to require the disclosure of customers' communications or records" [113]. Addressing some of the challenges posed by the outdated nature of ECPA, Congress introduced the LEADS Act in 2015, which covers non-US territory. Thus, the Act "would authorize the government to obtain the contents of electronic communications regardless of where those contents are stored if the account holder is a U.S. person" [113].

This feature addresses the practice of customer data storage in scattered overseas locations, and often not in the same country as the user. More recently, Microsoft was required to divulge content of a European subscriber's email, located on servers in Dublin, Ireland. If disclosing this content violates the laws of the foreign country, as Microsoft has argued, interpretation of the LEADS Act is ambiguous. Moreover, this opens up the possibility of nefarious foreign clients of US providers being shielded from US law. On the other hand, if this process were overturned, then non-US firms located on US soil are obliged to share data collected on US citizens with their foreign governments. So, for example, if the Chinese firm Alibaba located in Silicon Valley were asked to divulge information on US residents to the Chinese government, it would be compelled to oblige.

The FTC has created the principle of responsible use in its oversight of data collection, analysis and use. Regulation and the law are not going to effectively and efficiently address the problem of privacy and data misuse. Ultimately, cooperation, trust and responsibility will emerge from the transparency engendered by connectivity. Litigation or anonymity is not welfare enhancing for individuals or groups when there are cooperative options. Self-regulatory groups such as I am the Cavalry will sprout up and monitor data use. This group is a worldwide organization focused on the intersection of cybersecurity and human safety. Their focus is on the IoT or the Internet connection enabled by medical devices, home electronics, automobiles and public infrastructure. As the reach of digital technology extends beyond the reach of a society's ability to comprehend the risks, this group of researchers collects and analyzes data with a mission "to ensure that technologies with the potential to impact public safety and human life are worthy of our trust" [114].

Ownership of Domain Names

Currently, the National Telecommunications and Information Administration (NTIA) is working with ICANN and the international stakeholder community to determine the transition of stewardship of key Internet Domain name functions from the US to this international community. At issue is that some government or group of governments will exercise undue influence. Congress will make the case that the .mil and .gov domain names remain under the sole ownership and control of the US. The US is the only nation with a sponsored top-level domain (TLD) name for both its military and government (it does not need to be further defined by .us), a legacy of the important role played by the US Defense Department in the creation of the Internet. Similarly, the .edu TLD name is only granted to post-secondary institutions recognized by the US Department of Education [115].

This has national security and free speech implications because if certain domain names are captured by nations that wish to impose walls, then acquiring a particular domain name would give them that control. For example, the Chinese government recently issued legislation that would require all Internet domain names in China to be registered with a central domestic agency, which could impose further restrictions on these names. This action might tighten control over the Internet in China since non-Chinese domain names can then be easily monitored and blocked. Moreover, firms that conduct business in China would have to comply by having .cn domain names which would make them subject to Chinese oversight and comply with forced data localization, which might impinge on the free flow of information, adding more control to China's great firewall [116].

History plays an important role in ownership questions. Domain names historically owned by some nations can begin to acquire a new significance under changing circumstances.

Net Neutrality

Net neutrality has surfaced as a major policy issue in recent years. The forerunner of this issue is the idea of Common Carriage: in a town with one boat to ferry passengers across the river, the single boatman cannot charge the butcher more than he charges the carpenter for carrying passengers. Similarly, net neutrality posits that there can be no fast lanes or tollbooths to accelerate data packet transmission across Internet pipelines. From a historical perspective, the relevant framework was set by the 1934 Communications Act, signed by President Franklin Roosevelt, which aimed to treat radio and wire communication technology the same as railways and interstate commerce. The revised version was the 1996 Telecommunications Act which had the intent to deregulate as it required incumbent telecom carriers to provide interconnectedness or access to the network so that regional companies could enter the long-distance market.

The problem with this regulation was technology itself. The regulation focused on intra-modal technologies rather than inter-modal ones with technologies that provided the same function being treated differently. This dilemma is similar to that faced by the financial industry when regulation is by institution or historical product classification rather than across functional lines. Applying this idea to DCT the implication would be that all content transmitted across broadband pipelines should be treated the same and be subject to the same rules. Note that broadband refers to all digital transmission technologies, whether wireless, wire line, etc. However, in addition to telecommunications, broadband companies also process and store data, and offer security and spam protection.

There is reason to think the lines between wireless and wire line are beginning to blur, as mobile broadband service is becoming an increasingly effective substitute for wired broadband. When that happens fully, any wireless company will be able to offer a package of wireless phone, broadband, television and home phone service – putting companies like ATT in the same businesses as cable companies like Comcast. At the same time, cable companies are inching closer to the wireless business. Comcast, the biggest cable and wire line broadband provider, already offers home phone service over its cable network. It is aiming to build a nationwide Wi-Fi network that could allow it, with the help of additional access points, to provide national mobile phone service.

On June 14, 2016 US Court of Appeals for the D.C. Circuit has voted 2-1 to uphold FCC rules regulating ISPs under Title II of the 1984 Telecommunications Act, thereby classifying the Internet as a vital communication platform that must be treated as a common network much as the telephone system in the past. Under these rules, carriers cannot selectively block or speed up Internet traffic to consumers [117].

Underscoring the importance of the Internet to economic and social life, the decision written by Judges Davit Tatel and Sri Srinivasan stated that

> Indeed, given the tremendous impact third-party [I]nternet content has had on our society, it would be hard to deny its dominance in the broadband experience. Over the past two decades, this content has transformed nearly every aspect of our lives, from profound actions like choosing a leader, building a career, and falling in love to more quotidian ones like hailing a cab and watching a movie. The same assuredly cannot be said for broadband providers' own add-on applications. [118]

GATEKEEPERS

An economic landscape dominated by few firms has traditionally led to concerns about market pricing power. Monopolies can set price above cost and earn above normal profits. Economic power can be transformed into a form of influence whose mechanics are themselves not well understood. As we consume content on these large platforms such as Facebook, we react with comments and by sharing these comments with friends. Our reaction leaves data footprints which are assessed by computer programs. Algorithms then evaluate our reaction and update the information being fed to us. These are complex algorithms that interact with the data in a technology called machine learning. How updated content influences subsequent interactions is unclear, but there will be echoing effects whose final outcome is unknown. What is known is that these algorithms act as gatekeepers of information and knowledge for society and, as such, they wield enormous power. The point to note, however, is that no one group or person is exercising this power [119]. It is the very opacity of the binary code that is concerning, as it carries us into the realm of Ray Kurzweil's "singularity," where the binary world supersedes the world of living neurons [120].

The appropriate larger concern with OB is not pricing power but rather the power to shape ideas. The forces at work are subtle, and unsuspecting consumers become molded in a caricature of the homogenous user. The issue is not just point-in-time pricing power but rather the impact over time. By changing the context and environment or rules of the game, the game itself is changed. When competition for scarce resources drives each node to work harder to capture its share, others are strengthened by this very competition since survival demands a clever response.[2] Lack of competitors can stifle innovation by removing this struggle for survival, but when there is the *threat* of entry on the horizon even OBs are competing. However, strategic actions taken in the process of competition give OBs the power to influence popular culture. Voices raised on social media are echoed across nations in easy to remember, and replicate, sound bites, which then drown out complex analyses.[3]

In a powerful summation of the antecedents of the vote in Britain to leave the European Union, Tony Blair writes that "The Center must hold" not only in Britain but also globally. The tension that led to this referendum is universal. Populist insurgencies against the elite, fueled by movements of both the left and right,

> can spread and grow at scale and speed. Today's polarized and fragmented news coverage only encourages such insurgencies – an effect magnified many times by the *social media revolution*...the spirit of insurgency, the venting of anger at those in power and the addiction to simple, demagogic answers to complex problems are the same for both extremes. Underlying it all is a shared hostility to globalization. [122]

MY TAKE

What are the core values of the Internet and how do we protect them? DCT has precipitated adaptation and innovation on the inner margins, small and incremental. Past decisions have lingering effects whose implications unfold over time and impact the current environment, a phenomenon known as economic hysteresis [123]. The implication of this is waiting before acting. As more information is revealed over time, irreversible decisions can be made more effectively since some of the uncertainty is resolved. Adaptation and innovation itself present further challenges that must be responded to, in a cycle of learning and growth or competitive hysteresis.

So innovation from the edges, as Shane Greenstein [124] calls it, is a sharp survival tool. Rather than responding with massive changes, economic agents are taking baby steps as we ponder what values we want to protect and how we want to do so. How open do we want the Internet to be and how much autonomy do we want to relinquish to individual economic agents? A fragmented network economy with no coherent structure is self-destructing. Therefore, what kind of trade-off are we prepared to make between control and cohesion?

Notes

1. The connection between basic science and national security was crystalized in Vannevar Bush's report in 1945, after which Congress created the National Science Foundation. In 1958, the military-academic collaboration spurred by President Eisenhower culminated in the Advanced Research Projects Agency (ARPA), part of the Pentagon. The ARPANET is the network created by their mission of connecting computers to enable information transfer. Multiple smaller networks developed, but were not interconnected. In 1973, Robert Kahn conceived the idea of connecting networks into an inter-network, which subsequently became the Internet. With a capital I, the Internet now represents the entire technological ecosystem associated with this network.
2. This is the Red Queen Effect, which I discuss in the conclusion.
3. Facebook announced on June 29, 2016 that it would emphasize news about family and friends over more general news. They wrote on their website, "We are not in the business of picking which issues the world should read about. We are in the business of connecting people and ideas – and matching people with the stories they find most meaningful. Our integrity depends on being inclusive of all perspectives and view points, and using ranking to connect people with the stories and sources they find the most meaningful and engaging. We don't favor specific kinds of sources – or ideas. Our aim is to deliver the types of stories we've gotten feedback that an individual most wants to see" [121].

Bibliography

[112] Privacy Rights Clearinghouse. "Data Brokers and 'People Search' Sites." https://www.privacyrights.org/content/data-brokers-and-your-privacy. Accessed on June 17, 2016.

[113] Congressional Research Service. "Stored Communications Act: Reform of the Electronic Communications Privacy Act." https://www.fas.org/sgp/crs/misc/R44036.pdf Accessed on June 17, 2016.

[114]   I Am The Cavalry Cyber Safety Outreach. https://www.iamthecavalry. org. Accessed on June 17, 2016.

[115]   Schaefer, Brett, and Paul Rosenzweig. "How Congress Can Halt Government Control of the Internet." *The Daily Signal*, July 30, 2015. Accessed on June 14, 2016 from http://dailysignal.com/2015/07/30/ how-congress-can-halt-government-control-of-the-internet/

[116]   Sepulveda, Daniel, and Lawrence Strickling. (May 16, 2016), "China's Internet Domain Name Measure and the Digital Economy." Accessed on June 21, 2016 from https://blogs.state.gov/stories/2016/05/16/ china-s-internet-domain-name-measures-and-digital-economy

[117]   Byers, Alex. "Court Upholds Obama Backed Net Neutrality Rules." *Politico*. Accessed on June 15, 2016 from http://www.politico.com/ story/2016/06/court-upholds-obama-backed-net-neutrality-rules-224309

[118]   Statement made by D.C. Court of Appeals. Accessed on June 15, 2016 from https://www.cadc.uscourts.gov/internet/opinions.nsf/ 3F95E49183E6F8AF85257FD200505A3A/$file/15-1063-1619173. pdf

[119]   Tufekci, Zeynep. "The Real Bias Built In at Facebook." New York Times, May 19, 2016. Accessed on June 20, 2016 from http://www.nytimes. com/2016/05/19/opinion/the-real-bias-built-in-at-facebook.html? rref=collection%2Fcolumn%2Fzeynep-tufekci&action=click &contentCollection=opinion&region=stream&module=stream_uni t&version=latest&contentPlacement=1&pgtype=collection&_r=0

[120]   Kurzweil, Ray. *The Singularity is Near: When Humans Transcend Biology*. New York: Vintage, 2005.

[121]   Mosseri, Adam. "Building a Better News Feed for You." *Facebook Newsroom*. Accessed on July 1, 2016 from http://newsroom.fb.com/ news/2016/06/building-a-better-news-feed-for-you/

[122]   Blair, Tony. "Brexit's Stunning Coup," NYT OP Ed 6/25/16. Accessed on June 26, 2016 from Tony Blair, http://www.nytimes.com/2016/ 06/26/opinion/tony-blair-brexits-stunning-coup.html?action=click&pg type=Homepage&clickSource=story-heading&module=span-abc- region&region=span-abc-region&WT.nav=span-abc-region&_r=0

[123]   Dixit, Avinash. "Investment and Hysteresis." *Journal of Economic Perspectives* 6, no. 1, 1992.

[124]   Greenstein, Shane. *How the Internet Became Commercial: Innovation, Privatization, and the Birth of a New Network*. Princeton, NJ: Princeton University Press, 2016.

CHAPTER 9

The Conclusion

We Cooperate to Better Comprehend

Abstract We are in a moment of increasing interdependency because of our connections. Competition in terms of a zero-sum game is simply not an option. There is an acknowledgement, by recognizing our intertwined lives, that cooperating is the individually rational way forward. Since common knowledge is endemic in the network economy – my strategy choices are known almost before I know them – an open conversation about the game, or cooperation, is the best strategy. When the dominant pillars of the network economy are technology and human behavior, and technology has outrun the limits of the law and our ability to grasp its global outcomes, we adapt to fit the environment as we transform it. We cooperate in order to better comprehend.

Keywords Interdependency · Human behavior and law · Red Queen Effect · Consensus

In the network economy, connections are the primary mechanism for information sharing which *automatically* leads to informative prices and transparency. Whereas competition in the traditional model led to informative prices as diverse market participants interacted in a struggle for survival in markets with scarce resources, competition is transparency in a network economy.[1] Transparency reveals the benefits of pooling individual information in a cooperative model in order to create common

© The Author(s) 2017
S. Bhatt, *How Digital Communication Technology Shapes Markets*,
Palgrave Advances in the Economics of Innovation and Technology,
DOI 10.1007/978-3-319-47250-8_9

143

resources. Competition is not a static survival game, but rather a dynamic, ongoing adaptation to the new technology. The question posed in Chapter 1 – does the Internet move markets toward more competition or more cooperation – is best answered by recognizing this view of competition. Connectivity drives cooperative information gathering and sharing which then leads to granularity.

Bargaining is a form of cooperation. While there is a large literature focusing on bargaining in a zero-sum game, I suggest that in a network economy, cooperative bargaining is the rational outcome. Dixit and Skeath show that

> negotiations between buyers and sellers proceed to secure mutually advantageous trades. [Furthermore,] coalitions can get together to work out tentative deals as the individual people and groups continue the search for better alternatives. The process of deal making stops only when no person or coalition can negotiate anything better for itself [125].

We are in a moment of increasing interdependency because of our connections. Competition in terms of a zero-sum game is simply not an option. There is an acknowledgement, by recognizing our intertwined lives, that forming coalitions on a cooperative basis is the individually rational way. Common knowledge is endemic in the network economy – my strategy choices are known almost before I know them, so an open conversation about the game, or cooperation, is the best strategy. Human behavior and technology mutually adapt and reinforce each other in a dynamic interaction. Embedded in the Internet are implicit values that have bearing on US constitutional values of privacy, free speech and equal access. So, for example, algorithmic data collection, which creates personal stories by connecting the dots of information, could threaten commonly held notions of privacy. Economic agents who undertake these actions are embodying some social norms about privacy and free speech. Digital technology itself alter these values and norms, so we need to monitor these norms to ensure integrity of the First and Fourth Amendments. We need to create the requisite institutions. Consequently, cooperation is the *only* strategy amicable with technological capabilities and the law.

The three pillars of the network economy have become, effectively, technology, human behavior and the law. All three are in an inextricable dance of reinforcement and change, but the first pillar, technology, is the most rapidly evolving. Law is not only sluggish, but also reactive rather than proactive so it is never the first mover. The dominance of technology creates an imperative for adapting human behavior to a model of cooperation. We cooperate and share, not out of altruism but rather because it is the only strategy that is preferred in terms of individual goals and objectives.

Lessig writes,

> We believe that there are collective values that ought to regulate private action. ("Collective" just in the sense that all individuals acting alone will produce less of that value than if that individual action could be coordinated.) We are also committed to the idea that collective values should regulate the emerging technical world [18].

Cyberspace is beyond any particular jurisdiction, and there is a "shared community of interests that reaches beyond diplomatic ties into the hearts of ordinary citizens" [18].

The Red Queen

I posit that connectivity will subject the network economy to the Red Queen Effect. The game metaphor comes from Lewis Carroll's *Through the Looking-Glass* where Alice, after running with the Red Queen, finds herself in the same place where they began. She complains that in her country you would get somewhere when you ran fast, to which the Queen replies, "Now, here, you see, it takes all the running you can do, to keep in the same place" [126]. Ian Morris interprets the Red Queen Effect as a feedback process as

> species that evolve to fit better with their environment simultaneously transform that environment [so that] the race between values and environments is played out in billions of little cultural competitions, as individuals decide what is the right thing to do [127].

Red Queen competition, according to Bill Barnett, is the process of organizations adapting to and learning from challenges posed by their rivals and the environment, as all firms cope with scarce resources in a survival game [128]. In response to perceived threats, economic units, which are firms and consumers, enhance their capabilities, which sow the seeds for further challenges, eliciting renewed creativity and so on in a virtuous cycle.

The unique perspective to the Red Queen Effect in a digital world is that competition transcends the survival game, as information sharing or pooling of resources in a cooperative world generates higher survival probabilities. The whole is more than the sum of the parts so firms cooperate to transform the environment or enlarge the size of the pie instead of merely competing over its subdivision. For instance, the consumptions space has altered as individuals prefer human connections over material possessions, as in temporary ownership and instant access through rental contracts and streaming entertainment.

My Take

To understand the implications of connectivity we need to grasp what its absence means. It would mean isolation, anonymity and obscurity. Consider the following question: if a tree falls in the forest and no one hears it, does it make a sound? In the same vein, if one's very presence is unknown or not acknowledged, do we exist? Connections provide context for our lives and legitimate our existence. And so we share in a quest for identity and confirmation of membership in a social group. Digital technology has contextualized us in a universal social group that spans the globe.

Along the spectrum of anonymity and identity is also the parallel line of control. With anonymity comes complete freedom while with identity in cyberspace, there is potential for control. While identity facilitates commerce, it can also be used for centralized authority. At the other extreme, anonymity not only permits free speech, but also enables malignant behavior and could end up destroying the social fabric of the community.

Fukuyama makes the case that never in history were humans isolated. Identity politics was based on recognition of one's identity and a desire for the freedom "to not be ruled by those who are inferior or less worthy" and to govern one's own group. Nation building then became a search for identity [129]. In a connected world, as our identities become intertwined, perhaps the concept of nation itself will become amorphous. With a global identity, global institutions then arise to form the core of the global state itself. The Internet is one of them.

The other global institution is social capital, which undergirds the digital economy. Connectivity is taking us into human socialization beyond kinship ties, religion and national identity. Culture determines which ties are formed and whether they are strong or weak and directional or not and thereby creates social capital. This strong global institution is a culture of TRR&R among individuals. The global democratization of culture, made possible by DCT, undergirds the digital economy.

These global institutions are the pillars upon which the worldwide network economy rests.

We are moving to a social structure, economic and political institutions, culture and values which is similar in its manifestation to the forager society where equality trumped hierarchy. Morris writes,

> Foraging groups sometimes have to make important collective decisions, particularly about where to move next in the endless quest for food, but most groups have developed methods that make it difficult for one person or even one small group to seize control of the decision-making process. The most popular solution is to discuss every decision over and over again in subgroups, until a consensus begins to take shape, and at that point, even the strongest-willed dissenters tend to turn into yes-men and get on board with majority opinion [127].

Cooperation and consensus prevail today, because the alternative would be mutually destructive. Technology has outrun the limits of the law and our ability to comprehend its global outcomes. We cooperate in order to better comprehend.

NOTE

1. Non-disclosure agreements in many industries, particularly the technology industry, would appear to limit transparency and information sharing. However, these contracts suggest short run strategic imperatives for the firms themselves rather than conferring longer-term advantages.

BIBLIOGRAPHY

[125] Dixit, Avinash, Susan Skeath, and David Reiley. *Games of Strategy*, 3rd ed. New York: W.W. Norton, 2009.

[126] Carroll, Lewis. *Alice Through the Looking-Glass*, 1872. Limpsfield and London: Dragon's World, 1989.

[127] Morris, Ian. *Foragers, Farmers and Fossil Fuels: How Human Values Evolve*. Princeton, NJ: Princeton University Press, 2015.

[128]   Barnett, William. *The Red Queen among Organizations: How Competitiveness Evolves.* Princeton, NJ: Princeton University Press, 2008.

[129]   Fukuyama, Francis. *The Origins of Political Order: From Prehuman Times to the French Revolution.* New York: Farrar, Straus and Giroux, 2012.

INDEX

A

Advertising, 22, 33, 51, 72, 73, 74, 78, 82, 86, 87, 94, 95, 96, 97, 98

Amazon, 4, 14, 25, 47, 48, 50, 51, 52, 77, 78, 79, 83, 129, 135

Apple, 8, 12, 18, 31, 32, 34, 48, 50, 63, 78, 82, 83, 100, 108, 121, 127, 135

Architecture, 10, 24, 25, 68, 116

B

Big data, 7, 9, 20, 135

Broadband, 3, 81, 84, 85

C

Capital, 8, 9, 14, 21, 22, 37, 40, 44, 67, 146, 147

CIME, 10, 65, 72, 86

Commercial data brokers (CDB), 135

Communication, 2, 8, 10, 25, 63, 65, 66, 71, 84, 86, 87, 92, 93, 94, 122, 125, 127, 136

Connectivity, 2, 3, 8, 17, 24, 37, 107, 111, 122, 145, 146

Cooperation, 2, 3, 6, 10, 22, 24, 29, 32, 111, 119, 136, 144, 147

Coordination, 3, 24, 32, 33, 37, 50, 107, 109, 111, 114, 115

Copying model, 113

D

Delay hypothesis, 18

Digital Communication Technology (DCT), 2, 10, 11, 58

E

Economies of Scale, 50

Encryption, 127

Entrepreneurship, 3, 37, 57

F

Facebook, 4, 8, 22, 40, 50, 52, 76, 78, 86, 87, 88, 89, 94, 121, 128, 135, 139

© The Author(s) 2017
S. Bhatt, *How Digital Communication Technology Shapes Markets,*
Palgrave Advances in the Economics of Innovation and Technology,
DOI 10.1007/978-3-319-47250-8

G

Gatekeeper, 12, 34, 95,
     128, 139
General Purpose Technology, 24, 25
Google, 4, 8, 12, 14, 22, 50, 76, 79,
     85, 94, 116, 128, 135
Granular, 2, 11, 19, 31, 48, 52, 67,
     87, 111
Graph Theory, 11

H

Homophily, 105
Hubs, 47, 115, 116
Hysteresis, 140

I

Independent contractor, 57
Industrial revolution, 8, 18, 59
Information cascade, 105, 111
In-links, 115
Innovation, 20, 21, 24, 25, 26, 53, 62,
     63, 67, 93, 114, 135, 140
Intermediary, 7, 21, 31, 34, 35, 73,
     76, 78, 83, 90, 109
Internet, 6, 7, 8, 10, 11, 24, 25, 30,
     31, 33, 34, 35, 36, 37, 40, 49, 52,
     64, 73, 74, 76, 79, 80, 81, 82, 84,
     85, 88, 89, 92, 98, 106, 116,
     124, 126, 133–141,
     144, 146
Internet of Things, 10, 135
Investment function, 110
IPad, 58, 64, 83, 95, 114
IPhone, 8, 18, 51, 58, 66, 95,
     97, 114
IPO, 40

L

Liquidity function, 109, 111

M

Markets, 4, 5, 6, 7, 8, 10, 18, 19, 32,
     33, 34, 35, 40, 51, 52, 58, 59, 61,
     63, 84, 107, 113, 115, 123,
     143, 144
Mobile, 3, 7, 8, 9, 10, 25, 49, 50, 78,
     82, 84, 86, 87, 90, 93, 95–96,
     110, 129
Multi-homing, 77, 78, 82
Multi-sided market, 32–34

N

National Security Agency, 126
Network economy, 2, 3, 6, 7, 10, 18,
     20, 29, 30, 35, 36, 51, 75, 76,
     129, 141, 143, 144, 145
Node, 10, 11, 12, 14, 30, 34, 106,
     111, 115, 116, 117, 133, 140

O

On-demand, 57, 60, 71, 80, 81
Organizational behemoth, 50
OTT, 76, 80, 81, 82, 87, 128
Ownership, 2, 7, 9, 23, 26, 51, 59, 60,
     65, 72, 76, 79, 80, 82, 99, 101,
     110, 123, 134, 135,
     137, 146

P

Payments function, 108, 109
Peer-to-peer, 59, 99

Power Law, 115
Privacy, 9, 10, 11, 20, 21, 23, 26, 92,
    96, 119, 120, 121, 122, 123,
    124, 126

R
Red Queen effect, 145

S
Savings function, 108
Shock, 2
Social capital, 22, 146–147
Social Media, 86
Social network, 9, 50, 67, 86, 89, 124
Startup, 37, 40, 44, 48

T
Trust, responsibility and rights, 107

U
Uber, 3, 25, 31, 33, 34, 48, 49, 59, 100
Unbundling/unbundled, 2, 8, 48, 49,
    66, 75, 76, 79, 81, 86, 98, 107,
    109, 110

W
Word-of-mouth, 35, 87

Y
Yelp, 35, 100